PROCEEDINGS

OF THE

ESSEX INSTITUTE:

1864.

MONDAY, JANUARY 11. Evening meeting.

The President, A. Huntington, in the chair.

Donations to the Library and Cabinets were announced.

Rev. G. W. Briggs occupied the evening in reading a portion of a Memoir of the late President of the Institute, Hon. D. A. White.

Adjourned to Thursday evening next, for the continuation of the reading of the Memoir.

WEDNESDAY, JANUARY 13. Ordinary meeting.

J. G. Waters, in the chair.

E. K. Roberts was appointed Secretary pro tempore.

Arthur Kemble, and William Neilson of Salem, were elected Resident Members; Benjamin Peirce of Cambridge, James B. Endicott now in England, and William Endicott now in China, Corresponding Members.

THURSDAY, JANUARY 14. Adjourned evening meeting.

The President in the chair.

H. M. Brooks was elected Secretary pro tempore.

Rev. Dr. Briggs finished the reading of his Memoir of the late Judge White.

The thanks of the Institute were voted to Rev. Dr. Briggs, for his valuable and interesting Memoir of our late President, and a copy was requested for publication. (See Historical Collections, VI, No. I.)

MONDAY, JANUARY 25. Evening meeting.

The President in the chair.

Donations were announced to the Library and Cabinets.

Letters were read, from G. A. Ward accepting membership; from Corporation of Yale College; Trustees of the Newburyport Public Library; and New Haven Colony Historical Society, acknowledging the receipt of Publications: from R. S. Rantoul, in relation to the naming of Forts in Marblehead and Gloucester.

George A. Ward read a communication, giving an account of the formation of the ESSEX HISTORICAL SOCIETY, forty-two years ago last June.

Allusions having been made in Mr. Ward's communication, to the existence of the frame of the original "First Church," in Salem, on the land of David Nichols, rear of Boston street, considerable discussion ensued, as to the proof of the above mentioned frame being that of the "First Church." The President, Francis Peabody, G. A. Ward, A. C. Goodell Jr., and Rev. G. D. Wildes participated in the discussion; the arguments adduced seemed to favor the affimative of the question.

The thanks of the Institute were voted to Mr. Ward, for his valuable communication and a copy was requested for publication. (See Historical Collections, VI, No I.)

MONDAY, FEBRUARY 8. Evening meeting.

The President in the chair.

Donations to the Library and Cabinets were announced.

Letters were read, from Wm. Neilson accepting membership:

from the Smithsonian Institution, Washington, acknowledging the receipt of Publications: from Jonathan Pearson of Schenectady, in relation to the publications.

The Secretary read a communication from D. M. Balch, "*On the Sodalite at Salem.*" Referred to the committee on publications.

F. W. Putnam read a communication from George H. Emerson of Cambridge, "*On Magnetite, and an Unknown Mineral at Nahant.*" Referred to the publication committee.

Rev. G. D. Wildes spoke of the thoroughly English aspects of several of our olden towns in the County of Essex, noting particularly those of Ipswich, as illustrating to the untravelled eye, the marked features of the English rural town. Probably no County in the State in its local names and physical character is more suggestive of associations connected with the mother land.

A. C. Goodell Jr., in presenting to the meeting, one of the parts, (viz: the deed to the grantees, Edward Winslow and Robert Cushman,) of the original indenture or patent from Lord Sheffield, of the territory of Cape Ann, which indenture was deposited in the archieves of the Institute by J. Wingate Thornton Esq., of Boston, gave a brief account of the dates of the several voyages of discovery, charters and settlements by Englishmen in America; and specially referred to the earlier grants and charters of the planters at New Plymouth and Massachusetts Bay.

The instrument deposited by Mr. Thornton, bearing date Jan. 1, 1623–4, he declared to be the grant under which the New Plymouth people first laid claim to Cape Ann, and began that series of settlements by fishermen and planters which laid the foundation of this flourishing Commonwealth.

Mr. Wildes followed Mr. Goodell in some remarks as to the great value of such documents, and alluded to the care taken of similar articles in the British Museum, mentioning several very valuable historical relics which he had seen in that collection.

The thanks of the Institute were voted to Mr. Thornton for this valuable contribution.

WEDNESDAY, FEBRUARY 10. Ordinary meeting.
H. J. Cross in the chair.

Charles Creesey and Joshua Safford, of Salem, were elected Resident Members.

MONDAY, FEBRUARY 22. Evening meeting.
Vice President, A. C. Goodell Jr., in the chair.

Donations to the Library and Cabinets were announced.

F. W. Putnam presented a communication by A. S. Packard Jr., of Brunswick, Maine, entitled "*Notes on the Family Zygænidæ.*" Referred to the Committee on Publications.

R. S. Rantoul read the following communications which he had recently received from the War Department, at Washington, accompanying the same with a brief account of his visit to Washington and his interview with Mr. Whiting, the Solicitor for the Department, in relation to the subject of naming the Forts in Gloucester and Marblehead:—

WAR DEPARTMENT,
Washington City, Feb. 8th, 1864.

ROBERT S. RANTOUL Esq.,

DEAR SIR,

I have the pleasure of enclosing the order of the Secretary of War made at my request in accordance with the wishes of the Essex Institute, naming Fort Glover and Fort Conant.

Respectfully, your obedient servant,

WILLIAM WHITING,
Solicitor of the War Department.

WAR DEPARTMENT,
Washington City, Feb. 7th, 1864.

SIR,

The Secretary of War directs me to acknowledge the

receipt of Mr. Robert S. Rantoul's communication dated January 22d, addressed to you and enclosing copy of a resolution passed by the "Essex Institute" of Salem, Massachusetts, recommending that the fortifications now erecting in Marblehead be named "Fort Glover," and the works designated for the "Stage" in Gloucester "Fort Conant."

In reply, I am instructed to inform you, that the Secretary regards the names proposed as suitable designations of these defences, and that he has ordered that they be named accordingly.

Very respectfully, your obedient servant,

ED. M. CAMBY,
Brigadier General, A. A. G.

Hon. WM. WHITING,
Solicitor of the War Department.

The chair remarked as follows: soon after the publication of Mr. W. P. Upham's Memoir of Gen. John Glover of Marblehead, S. H. Phillips Esq., suggested the propriety of having one of the Forts about to be constructed in Marblehead named "Fort Glover."

At a meeting of the Essex Institute, held on Wednesday, Sept. 2, 1863, on motion of Mr. W. P. Upham, a committee, consisting of Messrs. W. P. Upham and A. C. Goodell Jr., was appointed to coöperate with the town authorities and citizens of Marblehead in such a manner as may be deemed appropriate to accomplish this object.

At a meeting held on Monday evening, Dec. 14, 1863, the subject of naming the fortifications designed for the "Stage" in Gloucester, "Fort Conant," suggested in a letter to Mr. Goodell by J. Wingate Thornton Esq., of Boston, in honor of Roger Conant, the founder of the first plantation in Massachusetts Bay, was brought to the notice of the Institute and referred to the same Committee who had under consideration the naming of the Fort at Marblehead.

Mr. Goodell moreover stated that as the business for which the Committee was appointed had been so fully accomplished by Mr. Rantoul, he desired to be excused from further duty.

This was voted, and also a resolution of thanks to Mr. Rantoul.

Adjourned to meet on Monday of next week, Feb. 29th, and voted that meetings be held on every Monday until otherwise ordered.

WEDNESDAY, FEBRUARY 24. Ordinary meeting.

H. F. King in the chair.

Henry R. Stone of Salem, was elected a Resident Member.

Charles E. Hamlin of Waterville, Me., and S. I. Smith of Norway, Me., were elected Corresponding Members.

MONDAY, FEBRUARY 29. Evening meeting.

Vice President, A. C. Goodell Jr., in the chair.

Donations were announced to the Library and Cabinets.

Letters were read, from Henry R. Stone, accepting membership: from B. F. Mudge, of Quindaro, Wyandote Co., Kansas, in relation to the Geological survey of that State: from Trustees of the New York State Library; Historical Society of Pennsylvania; Henry A. Smith of Cleveland, Ohio; J. Henry Stickney of Baltimore, Md.; and N. Paine of Worcester, relating to the publications.

F. W. Putnam made some remarks on Orthopterous Insects, suggested by specimens presented to the Institute by Miss Edmands.

Mr. Putnam presented a communication from A. E. Verrill of Cambridge, entitled "*Synopsis of the Polyps collected during the years* 1853–6, *by Dr. Wm. Stimpson, Naturalist to the North Pacific Expedition, commanded by Captains Ringgold and Rogers.*" Referred to the publication committee.

The Secretary read the following communication from Geo. A. Ward, in regard to the naming of "Fort Lee."

"While at work in reconstructing the fort on Salem Neck in 1812 as a member of the Salem Light Infantry, my grandfather informed me that it was originally planned by General CHARLES LEE, and that he gave instructions regarding it, and that his name was given to it. My said Grandfather was of the Committee of Safety and had considerable to do as to the fortifications in the neighborhood of Salem, and I think he could not be mistaken as to Fort Lee."

Extracts from the Town records were read in relation to this subject, and remarks were offered by the chair, H. Wheatland, W. P. Upham and others. Some suggesting that the Fort was named for Colonel W. R. Lee, formerly collector of Salem and an active officer in the Revolution.

The chair presented in behalf of J. V. Browne, a copper plate, on which was engraved the likeness of Rev. Joseph Sewall of Boston, and gave a brief sketch of the life of Mr. Sewall.

T. Ropes made some enquiries relative to the old Friends Meeting House, on the South side of Essex street, between Monroe and Dean streets, which were replied to by the chair.

John M. Ives spoke of the new silk worms that feed on the Ailanthus, and remarks were offered by F. W. Putnam and others on silk producing worms.

The remainder of the evening was occupied by F. W. Putnam, who gave a general view of the geological succession of animals, and their geographical distribution at the present time.

John H. Bettis and Robert Brookhouse 3d, of Salem, were elected Resident Members.

MONDAY, MARCH 7. Evening meeting.

Vice President, A. C. Goodell Jr., in the chair.

Donations to the Library and Cabinets were announced.

Letters were read, from C. E. Hamlin and S. I. Smith, accepting membership: from S. Jillson respecting some Birds.

H. Wheatland. read extracts from the Records of the Superior Court of Judicature and the Inferior Court of Common Pleas (1766) relative to one Jenny Slew of Ipswich, Spinster, (colored woman) vs. John Whipple Jr. of Ipswich, claiming damages for his detention of her as a slave. The judgment of the Inferior Court was reversed by the Higher Court and the plaintiff recovered her liberty and damages.

Rev. G. D. Wildes spoke of Domestic Servitude as it existed

in this country prior to the Revolution, and instanced the case of a Norwegian girl in his Grandfather's family, whose services were purchased for a term of years.

The chair alluded to a similar case in Manchester.

Mr. Wildes spoke of Marblehead as presenting a near and most interesting field of Antiquarian research for the younger members of the Institute, whose minds might be directed to that department. St. Michael's Church, with its ancient Church yard; the old mansions of that formerly flourishing seaport; the history of several families identified with Colonial and Revolutionary history, would be found to present most interesting points of enquiry.

Mr. Wildes also spoke of Christ Church, Cambridge as perhaps the best specimen, in this country; of the English Village Church of the last Century. It was a question whether the frame of this Church was brought from England or not. Mr. W. gave an interesting account of the Vassal family, in connection with this Church, and of the several old mansions, still marking the social life of Cambridge in the Ante-revolutionary history of the town. A visit to Cambridge, in connection with researches into the history of some of these, even now elegant residences of a later generation, would be found to be full of interest and instruction.

Remarks of a conversational character from Messrs. Wildes, Beaman, the chair and others, relating to Boston and its vicinity in Revolutionary times occupied the rest of the evening.

A Committee consisting of Messrs. F. W. Putnam, J. A. Gillis, R. S. Rantoul, W. P. Upham and H. Wheatland were appointed to revise the Constitution and By-Laws.

MONDAY, MARCH 14. Evening meeting.

Vice President, A. C. Goodell Jr., in the chair.

Donations to the Library and Cabinets were announced.

Letters were read from Joseph A. Goldthwait of New Berne, N. C., relating to specimens sent to the Institute: from Wm. A.

Smith of Worcester, Mrs. P. A. Hanaford of Beverly, S. D. Bell of Manchester, N. H., C. M. Tracy and F. E. Oliver of Lynn, relating to business matters.

The Secretary read some extracts from the Records of two Aqueduct Corporations, which, though limited in their operations, are interesting as relating to the history of the introduction of water into this city. (See Historical Collections, VI, No. I.)

F. W. Putnam exhibited the Pea Hen recently presented by F. Peabody and mounted by S. Jillson. This Hen had been kept on the grounds of Col. Peabody for seventeen years; about two years since she commenced to assume the plumage characteristic of the male, and had so far accomplished this object that at the time of her death she had attained the "train" and the brilliant colors of the male. Mr. Putnam stated that Latham, in his Synopsis of Birds, mentioned two such instances that had come under his observation. He also said that similar cases had been noticed among other birds, and was quite common in the English Pheasant. Similar changes in the external appearance were known to take place in some species of fishes.

A. C. Goodell Jr. read a portion of an account, presented by George B. Loring, of the houses on Essex street in 1793, written by Col. Pickman who died in 1819.

Thomas Morong of Gloucester, was elected a Resident Member. Jeremiah L. Hanaford of Watertown, and Benj. F. Mudge of Quindaro, Kansas, were elected Corresponding Members.

MONDAY, MARCH 21. Evening meeting.

Vice President, A. C. Goodell Jr., in the chair.

Donations to the Library were announced.

Letters were read, from the Trustees of the New York State Library, giving notice of the transmission of books: from Trustees of the Boston Public Library, acknowledging the receipt of publications: from Wm. A. Smith of Worcester, in relation to publications.

Mr. Goodell concluded the reading of Mr. Pickman's account

of the old houses on Essex street. Referred to the publication committee to be printed in the Historical Collections.

Some discussion followed relative to the old houses in Salem, participated in by Messrs. Ropes, Goodell and others.

MONDAY, MARCH 28. Evening meeting.

Vice President, A. C. Goodell Jr., in the chair.

Donations to the Library and Cabinet were announced.

F. W. Putnam made some remarks upon the Trilobites from the Braintree quarry, presented by A. S. Packard Jr.

The Rev. Mr. Wildes, presented to the Institute, several articles which he had procured in a visit to Newburyport this afternoon. One of these was a framed engraving of the body of Marshal Ney, as it appeared after being taken to a conventical house in Paris, immediately after his execution. The engraving, suppressed by the Allied Commander in the fear that it might tend to popular tumult, is supposed to be the only one in this country. It presents a most faithful portrait of the Marshal, and is not the least interesting among the historical objects in the collection of the Institute.

Mr. W. also presented to the Institute, on deposit, the bullet by which Capt Greenleaf was wounded in the fight with the Indians near Newbury, in 1695. It is hoped, that the buff coat, worn on the occasion by Capt. G., and still in the possession of his descendants, may eventually be obtained for the Institute.

A third article presented by Mr. W., on behalf of the Misses Tracey of Newburyport, was the snuff box of the eminent merchant Jeremiah Lee of Marblehead, the subject of the exquisite painting by Copley, now, with that of Madame Lee, in the possession of the Misses Tracey.

Another article presented by Mr. W., in behalf of E. W. Rand Esq. of Newburyport, was a pair of very ancient tongs, used for the purpose of lighting a pipe, and with various peculiar contrivances for securing reasonable comfort in smoking. Mr. W. accompanied the presentation with various interesting details as to these and other objects of interest, which might eventually be procured from the same sources for the collections of the Institute.

F. W. Putnam called the attention of the meeting to a singular monstrosity that had been presented by Mr. James Buffington of Salem. This was a young duck that had, apparently, an extra leg developed from its back. Upon dissection this leg proved to be made up in some parts, of two legs closely united. The portion joining the pelvis (the femur,) being single, but the second segment of the leg (tibia and fibula) was shortened and spread out, so as to allow the articulation of *two* tarso-metatarsal bones, and from this point the foot was nearly double, having six toes, the two small hind toes being wanting.

Mr. William Mansfield presented to the Institute a wooden model, used before the city government to illustrate the grade and direction of the proposed route of the Eastern Railroad, in 1837-8, through Washington Street, in Salem. This model contains, in miniature, all the buildings then standing on the land included in the present Washington Street South of Essex Street, except the "Marston building." The chair gave an historical sketch of these several buildings, and of earlier structures in the same locality.

Additions to the Museum and Library during January, February and March, 1864.

TO THE NATURAL HISTORY DEPARTMENT.

BERTRAM, JOHN. Specimen of Malachite.

BUFFINGTON, JAMES. Malformed Young Duck.

EDMANDS, MISS A. M. Collection of 24 species of North American Orthoptera, named by Mr. Scudder. 250 specimens of New England Spiders.

FELT, S. Q. Lime incrustation from Brazil.

GOLDTHWAIT, CAPT. J. A., New Berne, N.C. Fossil wood, a portion of a large tree, from Neuse River, near Kingston, N. C.

HAMLIN, PROF. C. E., Waterville, Me. 3 Salamanders, 2 species from Waterville.

HARRINGTON, CAPT. GEO. Fossil Shells from Gibraltar.

HARTT, C. F., St. John, N. B. Fossil coral, *Siderastrea siderea* Blainv. from Bermuda. 7 species of Minerals from Nova Scotia.

HARTT, J. W., St. John, N. B. Two specimens of Fossil Fish from the Albert Coal Mine, N. B.

KING, CAPT. H. F. Wood of the *Sophora Japonica.*

LYCEUM OF NATURAL HISTORY OF WILLIAMS COLLEGE (In exchange.) 19 species of Corals from Florida, named by Mr. Verrill.

MUSEUM OF COMP. ZOÖLOGY, Cambridge. (In exchange.) 32 species of corals from various localities. Named by Mr. Verrill. 16 specimens, 11 species of Bird's eggs from Florida, Grand Menan and Anticosti.

NEAL, JOS. Body of a Fox, for Skeleton.

ORDWAY, H. L., Ipswich. 34 specimens, 5 species of Spiders from Ipswich.

PACKARD JR., A. S., Brunswick, Me. Trilobites from the Braintree Quarry. 40 species, 200 specimens of Lepidoptera Maryland. 3 species, 8 specimens of Lepidoptera from England.

PEABODY, FRANCIS. A female Pea fowl which had assumed the characteristic plumage of the male.

PUTNAM, F. W. Iron Ore from Port Henry Mines, N. Y. Clay stones from Lake Champlain.

PUTNAM, CAPT. W. H. A. Copper from Chili. Dry shells and Echini from Caldera, Chili. Alcoholic specimens of Crabs and Starfishes from Caldera.

ROBINSON, JOHN. 40 specimens, 12 species of Insects from Salem.

RUSSELL, T. B. Geological specimen of Sand Stone.

SANBORN, F. G., Boston. 117 specimens, 28 species of Spiders from Essex County.

SMITH, LAWRENCE P. Insects from the Southern States.

SMITH, S. I., Norway, Me. Young Salamanders from Norway.

STEVENS, C. B. Skin of a Phatagin *Manis tetradactyla* from Madagascar.

STONE, W. H., Agent Port Henry Mines, N. Y. Specimen of the Cheever Iron Ore from four hundred feet depth.

TO THE HISTORICAL DEPARTMENT.

BARTON, WM. C. Model of a Chinese Vessel.

BROOKS, H. M. Rebel Prayer Book, from a Blockade Runner; Child's book of 1812.

BROWN JR., BENJ. Two United States Buttons.

BROWN, HORACE. Feather Cape from India.

COLE, MRS. N. D. Bust of Alex. Hamilton, (in plaster, bronzed.)

CLOUTMAN, WILLIAM R. 3 China, 3 Japanese and 3 Russia Coins. Japanese Inkstand. Brick from Captain Kid's Fort, and a Stone from the grave of "Paul and Virginia."

GREENOUGH, W. An old Musket, taken from a Blockade Runner.

KIMBALL, CAPT. THOMAS. Model of a Catamaran used on the coast of Brazil.

MANSFIELD, WM. Model used to illustrate the route of the Salem Tunnel.

NICHOLS, MRS. ANDREW. Tile from the old Gov. Winslow house, Plymouth.

NICHOLS, C. F. Stone from the Hoosac Tunnel, half a mile from the entrance.

NICHOLS, CAPT. JAMES B., 24th Reg. Mass. Vols. Stone from an old Spanish Fort at St. Augustine, Fla.

ORDWAY, LIEUT. A., 24th Reg. Mass. Vols. Rebel Musket

taken at the Battle of Roanoke Island. Portion of Rebel Flag-staff at Washington, N. C.

RAND, E. W., Newburyport. Ancient Tongs used for lighting a pipe.

ROBINSON, JOHN. War Relics from New Berne. Chinese Playing Cards.

ROGERS, EWD. S. Old Tiles for Fireplace ornaments.

RUSSELL, A. B. Cane made with a jackknife.

SMITH, LAWRENCE P. Rebel Sword.

WARD, G. A. The Waistcoat worn by Capt. Jonathan Haraden during the Revolutionary war.

WARD, W. R. L., New York. Shaving from a wrought iron cannon, made at Falls Village, Conn.

WIGGIN, J. K., Boston. Sword blade, from the cargo of the Anglo-Rebel Blockade Runner "Minna."

WILDES, REV. G. D. (On deposit.) An engraving of the body of Marshal Ney, from a drawing made soon after his execution. The bullet taken from the body of Capt. Greenleaf, wounded in a fight with the Indians near Newbury in 1695. The Snuff Box of the eminent merchant Jeremiah Lee of Marblehead.

TO THE LIBRARY.

CHASE, GEORGE C. Friend's Review, 20 numbers.

COLE, MRS. N. D. Salem Gazette, 1863, 1 vol. folio; Boston Daily Evening Traveller 1863, 2 vols. folio.

DAVIS, CHARLES of Beverly. Files of Beverly Citizen, vols. 1 and 2, folio, 1850 to 1853.

DECOSTA, B. F. of Charlestown. Footprints of Miles Standish, by Rev. B. F. Decosta, 12mo, pamph. Charlestown, 1864.

DODGE, ALLEN W. of Hamilton. Cushing's Newburyport, 12mo, 1826; Report on Hoosac Tunnel, Feb. 1863, 8vo, pamph.

FOOTE. C. Files of several County Papers, for September, October, November, and December, 1863.

GILLIS, JAMES A. Massachusetts State Registers for 1856 and 1858, 2 vols. 8vo. 25 Pamphlets.

HANAFORD, MRS. P. A. of Beverly. Several numbers of the New Jerusalem Messenger.

HODSDON, JOHN L of Augusta, Me. Annual Rep. of Adj. Gen. of Maine for 1862, 1 vol. 8vo. Augusta, 1863.

HOLMES, JOHN C. 24th Annual Rep. of Supt. of Public Instruction of Michigan, 8vo. Lansing, 1862.

JOHNSON, A. B. of Utica, N. Y. Our Monetary Condition, by A. B. Johnson, 8vo, pamph. Utica, 1864.

JOHNSON JR., CAPT. DANIEL H. Enrollment List, 5th Dist. Mass., Nov. 1863. 4to.

JOHNSON, MRS. LUCY P. Independent for 1863, fol. New York.

KILBY, W. H. of Eastport, Me. Eighth Annual Rep. Sec'y of Maine Board of Agriculture, 8vo. Augusta 1863.

LANGWORTHY, ISAAC P. of Boston. Spirit of Missions, ten numbers. Am. Tract Soc. of New York Annual Reports 28, 30, 32, 33, 34. Am. Tract Society of Boston. Reports 11, 37, 61, 65, 67.

LORD, N. J. Boston Post, Dec. 1863 and Jan. 1854.

LORING, GEORGE B. Two manuscript volumes containing the Expenses of Salem from 1788 to 1802, kept by Benjamin Pickman.

MASSACHUSETTS SECRETARY OF STATE. Mass. Public Documents 1862, 3 vols. 8vo; Census of Mass., 1 vol. 8vo, 1863; 19th and 20th Registration Reports, 2 vols. 8vo; Acts and Resolves for 1863.

MOORE, GEO. H. of New York. The Treason of Chas. Lee, by G. H. Moore, 1 vol. 8vo. New York, 1860. Historical Notes on the Employment of Negroes in the American army of the Revolution, by George H. Moore, 8vo, pamph. New York, 1862.

NICHOLS, CHARLES F. Collection of Handbills, &c.

NICHOLS, GEORGE. Christian Inquirer for 1863, 1 vol. fol. New York.

Nichols, Henry P. Several pamphlets.

Oliver, H. K. 27 pamphlets, including Legislative Documents and Town Reports.

Paine, Nathaniel of Worcester. Worcester Directory for 1864. 12mo.

Putnam, Capt. George D. Regulations of the Army of the Confederate States, 12mo. Richmond, 1863.

Rantoul, R. S. Several pamphlets.

Short, Joseph. Shepard's Sound Believer, 12mo. Boston, 1762.

Sibley, John L. of Cambridge. 38th Annual Rep. of Pres. and Fellows of Harv. College, 8vo, pamph. Cambridge, 1864.

Stevens, C. B. Pilot from July to Dec. 1807 and part of 1809, 3 vols. folio. London.

Stone, E. M. of Providence, R. I. Report of the Ministry at Large, Jan. 24, 1862. 8vo, pamph. Providence, 1864.

Swett, S. of Boston. Original planning, &c. of Bunker Hill Monument, by S. Swett, 8vo, pamph. Albany, 1863.

Upton, George. Scientific American, several numbers.

Upton, James. Magazine of Horticulture, vols. 8 to 29, 22 vols, 8vo, Boston, 1842, &c.; Horticulturist for 1863, 1 vol., 8vo, New York, 1863; Littell's Living Age, vols 77, 78, 79, 3 vols, 8vo, Boston, 1863; Am. Bapt. Missionary Magazine, vol. 43, Boston, 1863; Nautical Magazine, vols. 2 to 6, 5 vols., 8vo, New York, 1855, &c.. Barry's Fruit Garden, 1 vol., 12mo, New York, 1851; Kenrick's Am. Orchardist, 3d edition, 1 vol., 12mo, Boston, 1841; Kenrick's Am. Orchardist, 7th edition, 1 vol., 12mo, Boston, 1844; Johnston's Agricultural Chemistry, 1 vol., 12mo, New York, 1844; Downing's Fruit and Fruit Trees, 1 vol., 12mo, New York, 1845; Field's Pear Culture, 1 vol., 12mo, New York, 1859; Liebig's Agricultural Chemistry, 1 vol, 12mo, Cambridge, 1842; Hoare on Vine Roots, 1 vol., 12mo, London, 1844; Hoare on Grape Vines, 1 vol., 12mo, Boston, 1840; Manning's Book of Fruits, 1 vol., 12mo, Salem, 1838; Jaques on Fruit Trees, 1 vol., 12mo,

Worcester, 1849; Thomas' Fruit Culturist, 1 vol., 12mo, Auburn, 1849; Lindley's Horticulture, 1 vol., 12mo, New York, 1841; Elliott's Fruit Grower's Guide, 1 vol., 12mo, New York, 1854; Peter Schlemihl in America, 1 vol., 12mo, Philadelphia, 1838; Wood's Modern Pilgrims, 2 vols, 12mo, Boston, 1855; Lester's Glory and Shame of England, 2 vols., 12mo, New York, 1841; Pamphlets, 23.

WARD JR., CHARLES. New York Journal of Commerce from July to Dec., 1863; Philadelphia Directory for 1861, 8vo; Bradbury's History of Kennebunk Port, 12mo, Kennebunk, 1837; Commercial Relations of United States, 5 vols., 4to, Washington, (Pub. Doc.); Handbook to the Museum of Phil. Acad. Nat. Sci., 12mo, pamph., Philadelphia, 1862.

WATERS, J. LINTON of Chicago, Ill. 15th Annual Rep. of Trade and Commerce of Chicago, for 1863, 8vo, pamph.

WHEATLAND, MRS. B. Boston Daily Transcript, July to Dec., 1863, 1 vol., folio.

WHEATLAND, STEPHEN G. 45 Pamphlets.

WYMAN, T. B. of Charlestown. Genealogy of the Hunt Family, 4to, Boston, 1862-3.

BY EXCHANGE.

AMERICAN ANTIQUARIAN SOCIETY. Proceedings at meeting, Oct. 21, 1863, 8vo, pamph., Boston, 1863.

BOSTON SOCIETY OF NATURAL HISTORY. Proceedings, vol. IX, Sig. 15, 16, 17, 18, 19; Journal, vol. VII, No. 4.

CANADIAN INSTITUTE at Toronto. The Canadian Journal for Jan. 1864.

CHICAGO HISTORICAL SOCIETY. Hibbard's Discourse, "A Spiritual ground of Hope for the salvation of the country," 8vo, pamph., Chicago, 1863; Several Reports of the Sanitary Commission of Illinois.

EDITORS. Historical Magazine for Jan. Feb. and March, 1864, New York, 1864.

EDITORS. The British American for January, February and March, 1864, 8vo. Toronto 1864.

IOWA STATE HISTORICAL SOCIETY. Iowa Legis. Doc.; House Journal, 1854, 1856, 1858, 3 vols. 8vo; Senate Journal, 1854, 1856, 1858, 3 vols., 8vo; House Journal, Extra Session, 1856, 1 vol., 8vo; Iowa Laws, 1848, 1856, 1860, 6 vols., 8vo; Iowa Legis. Doc. 1859-60, 1 vol., 8vo; Iowa Journal of Constitutional Con., 1857, 1 vol., 8vo; Iowa Constitutional Debates, 1857, 2 vols., 8vo; Iowa Census Returns, 1857, 1 vol., 8vo; 4th An. Report of Iowa State Agricultural Society, 1858, 1 vol., 8vo; 16 Miscellaneous pamphlets.

LONG ISLAND HISTORICAL SOCIETY. Clark's Onondaga, 2 vols., 8vo, Syracuse, 1849; Stiles' Supplement to Hist. and Geneal. of Ancient Windsor, 8vo, Albany, 1863; N. Y. State Agr. Soc. Trans. 1861, 8vo; Longworth's Directory of New York, 1840-1, 12mo; William's N. Y. An. Reg., 1831, 1832, 1833, 1834, 1836, 1837, 6 vols., 12mo; Brown's History of the Shakers, 12mo, Troy, 1862; 12 pamphlets.

MONTREAL SOCIETY OF NATURAL HISTORY. The Canadian Naturalist and Geologist, Dec. 1863.

NEW YORK STATE LIBRARY, TRUSTEES OF. Laws of New-York, session 1863, 1 vol. 8vo; Journal of Senate, session 1863, 1 vol. 8vo.; Documents, Senate, session 1863, 5 vols. 8vo; Journal Assembly, session 1863, 1 vol. 8vo.; Documents, Assembly, session 1863, 9 vols. 8vo; N. Y. State Agr. Soc. Trans. 1862, 1 vol. 8vo.; N. Y. Med. Soc. Trans. 1863, 1 vol. 8vo.; Am. Inst. Trans. 1862, 1 vol., 8vo.; 16th Annual Report on State Cabinet, pamph. 8vo.

PHILADELPHIA ACADEMY OF NATURAL SCIENCES. Proceedings for August, Sept., Oct., Nov., Dec., 1863.

PUBLISHERS. North American Review for Jan. 1864.

MONDAY, APRIL 4. Evening meeting.

The President in the chair.

Donations were announced to the Library and Cabinets.

Letters were read, from J. L. Hanaford of Watertown, accepting membership; and from Wm. Graves of Newburyport, on business matters.

Rev. G. D. Wildes gave an account of Queen Elizabeth's yacht, and showed, by a drawing on the black board, that in model and rigging it very nearly resembled the North River Sloops of the present day.

He suggested that an account of the different kind of vessels used from the early settlement of the country to the present day would be a valuable contribution to our commercial history.

A. C. Goodell Jr. called attention to the late discovery of a sunken vessel near Yarmouth on Cape Cod, supposed to have foundered there in 1623, which illustrated the manner of building at that period.

F. W. Putnam, in reply to questions, described the characteristic form of the breast bone of swimming birds and the different modes of progression among fishes. Mr. Putnam alluded to the erroneous views in regard to moths, as recently given in the newspapers and gave an account of the various species which are so destructive to furs, carpets, cloths, &c.

Voted; that the committee, appointed on the 7th of March, on the Constitution and By-Laws, be requested to nominate a list of officers for election at the annual meeting.

MONDAY, APRIL 11. Evening meeting.

Vice President, A. C. Goodell Jr., in the chair.

Adjourned to Monday Evening the 18th inst.

MONDAY, APRIL 18. Evening meeting.

Vice President, A. C. Goodell Jr., in the chair.

Donations to the Library and Cabinets announced.

Letters were read, from Thomas Morong of Lanesville and B. F. Mudge of Quindaro, Kansas, accepting membership; from Long Island Historical Society, Smithsonian Institution, and George A. Ward relating to books transmitted to the Library;

Messrs. Ticknor and Fields on business matters; from David Choate in reply to queries proposed; from Miss M. B. Derby accompanying a donation of a Burmese Idol sent from India in 1825, by her brother the late Capt. Alfred F. Derby; from James T. Tucker, of the staff of General Banks, relating to a donation to the Historical Department, of the envelope, franked by President Lincoln which enclosed his recognition of the election of Governer Hahn of Louisiana.

Rev. G. D. Wildes exhibited a piece of stone taken from a window sill in Kenilworth Castle, and made some interesting remarks about that celebrated place. Mr. W. also exhibited several views of the house in which Shakespeare was born and of other interesting localities in the vicinity of Stratford upon Avon, and gave a description of the same.

The remarks of Mr. Wildes called forth a general discussion upon the life and writings of Shakespeare.

F. W. Putnam mentioned that Mr. James H. Emerton had found a female Lump Fish *Cyclopterus lumpus*, having matured eggs, just on the point of being laid, and had made an estimate of their number, which amounted to 258,372. Five hundred eggs weighed 43 grains.

Col. J. H. Wildes, Asst. Surveyor General of California was elected a Corresponding Member.

Voted to adjourn to Monday evening, the 25th inst.

MONDAY, APRIL 25. Evening meeting.

The President in the chair.

Donations were announced to the Library and Cabinets. Letters were read, from Newburyport Public Library and Pennsylvania Historical Society, acknowledging the receipt of publications; from Messrs. Crosby & Nichols of Boston, and Henry A. Smith of Cleveland, Ohio, on business matters; from Mrs. P. A. Hanaford in relation to holding a Field Meeting in Reading.

F. W. Putnam, from the Committee on the Constitution and

By-Laws, submitted the first reading of the amendments to the Constitution to be acted upon at the annual meeting.

Mr. Putnam read a communication from A. S. Packard Jr. of Brunswick, Me., entitled "The Humble Bees of New England, and their parasites, with notices of a new species of Anthophorabia, and a new genus of Proctotrupidæ." Referred to the Publication Committee.

The subject which occupied a portion of the last meeting, and which had engrossed the attention of the Literary and Historical Societies during the past week, the ter-centenary birth day of Shakespeare, was resumed, remarks being made by the chair, Messrs. Wildes, Beaman and others.

John Kilburn of Salem was elected a Resident Member.

MONDAY MAY 2. Evening Meeting.

Vice President, A. C. Goodell Jr., in the chair.

Donations to the Cabinets and Library were announced.

Letters were read, from Henry Saltonstall of Boston, Justin Rideout of Boston, L. Saltonstall of Newton, and the Postmaster of Boston, on business matters; from the Misses Derby, relating to a donation of books to the Library; from the Mass. Historical Society, acknowledging the receipt of publications; from John Kilburn, accepting membership.

A variety of May-flowers having been placed upon the table, and the subject of May-day festivals having been alluded to, the Chair remarked that the return of another May-day, with its accompanying festivities, invites us to consider the pleasant change now working in the public mind of New England with regard to the observance of this ancient holiday of our Motherland.

The very name of May, not less than the practices used to usher in the month, runs back into the obscurity of antiquity. The poet Ovid, whose surmise has been generally adopted, derives it from the names of several Roman deities, among whom is the fair Maia, the mother of Mercury. But there are, on the other hand, some reasons to support the conjecture that the name is of Teutonic origin; and, as this conjecture neither wounds our vanity nor conflicts with history, we may safely assume it to be the true one, and so unbridle fancy to carry back our May-day festivals beyond the time of the Heptarchy, into the woods of Germany, aud among those hilarious wild-men, the primitive ancestors of our Saxon stock.

Whatever gave rise to the ceremonies of May-day—whether they are a relic of the early "mythology of the Teutonic peoples," or a continuation of the *Floralia* of the Romans, or a Christian festival in honor of the Blessed Virgin, as has been variously supposed by different investigators of the subject—all are agreed that, in England, at least, they are of so ancient observance that "the memory of man runneth not to the contrary;" and that, universally, they symbolize the joy of mankind at the triumph of the Sun over the frosts and barrenness of Winter.

The celebration of the May-games was extremely distasteful to the Puritans and other early reformers in the English Church; and, doubtless, the many excesses of the revellers—the wantonness and debauchery inseparable from these festivals—were sufficiently scandalous to all pious and moral men. Latimer, who suffered martyrdom in the reign of Mary, discloses another objection to these pastimes in a sermon preached before the young King Edward against the popular observance of Robin Hood's day, which, he complains, sometimes drew all the parish away from church. "I thought," he mournfully says, concerning an instance of this kind within his own experience, "my rochet would have been regarded; but it would not serve, it was faine to give place to Robin Hood's men."

The Puritans were certainly not steeled against all the sweet influences of nature, nor backward in their enjoyment and praise of the beauties of Spring; and it was the chief of Puritan poets whose "Song on May Morning," remains to this day unapproachable in its excellence.

But the Puritans were not blind to the evils already alluded to, and, moreover, it is clear that they considered the May-pole to be a relic of those heathen rites performed by the ancients in their worship of the goddess Flora: it was for this reason that Philip Stubs arraigned the May-games in 1595, in his "Anatomie of Abuses;" and for this reason sixty years later, Thomas Hall made them the subject of his "*Funebria Floræ*; or, Downfall of May-games," &c. Here, in New England, our good old Governor Bradford, of Plymouth, also condemned them for the same reason.

Not long after the landing of the Pilgrims at Plymouth some events occurred in their neighborhood, which called forth an official denunciation of May-day festivities, by the colonial authorities; and the rebuke was administered in so emphatic a manner that, if it has not effectually prevented a repetition of these ceremonies for all time, in New England, it has, at least, brought upon them a stigma which the lapse of two centuries has not wholly removed.

The Chair then proceeded to give an account of "Thomas Morton, of Clifford's Inn, Gent."—as he styles himself in his "New English Canaan"—and of the famous May-day revels at "Ma-re Mount," now Mount Wollaston, in Quincy, which were celebrated under his direction in 1626.

After detailing the particulars of the action of the colonial authorities against Morton, the dispersion of his followers and the destruction of his plantation, the Chair narrated the principal known facts of his subsequent career down to the time of his death at York in Maine, in 1646, and stated that this first May-day jubilee continued to be, for generations, the last. There had been May-day festivities in Maine before the affair at Mt. Wollaston, and there is some reason to suppose that Morton was a participant in those revelries; but, after his expulsion, and the destruction of his plantation at "Mount Dagon" no Puritan father was ever offended by the sight of the scandalous altar of Flora enticingly set up before the innocent eyes of his children. But the times are greatly changed since the dark and troubled days of the Pilgrims. There is now, happily, no need of ceaseless vigilance and the most sensitive jealousy in guarding a tender faith from the two-fold danger of relapsing into error or being contaminated by new and specious fallacies. Around our morals, our faith, our liberties, as their great bulwark of safety, modern science has thrown a network of invulnerable truths till old besetting evils have lost their power of harm forever.

No prejudices, then, based on the experience of an age remote and quite unlike the present, should be suffered to interfere with the celebration of the pleasant and pure festivities which of late years are beginning to be observed on May-day, in some parts of New England. It is to be hoped, rather, that we shall add some day in May to the list of legal holidays, and that, from the St. Croix to where "Mine Host of Ma-re Mount" sleeps under the brow of Agamenticus, and thence to Mount Wollaston, where he held his revels, and so along the entire boundary of our Union, May morning will evermore be held sacred to the celebration of the sun's return, the bursting of green buds and the birth of the flowers.

The wild flowers exhibited at the meeting, by those who went a-maying, were described by G. D. Phippen in the following manner:

Hepatica triloba, which differs but slightly from an anemone, is one of the earliest plants that has any pretensions to beauty, and is found in oaken woods, peeping up among the dried leaves, in close proximity with drifts of snow. It was mentioned by Higginson in 1629, and described by Josselyn in his New England Rarities printed in 1672, as "Noble Liverwort, one sort with white flowers and the other with blew." The Rev. Dr. Cutler mentions it in 1784, and Collinson writes to Bartram of Philadelphia in 1739, that "out of some mould sent with other plants has come up your Hepatica."

Anemone nemorosa, or Wind Flower. This little flower or

a co-species was described centuries ago by Pliny, and long before Gerard wrote "That this floure doth neuer open itself but when the wind doth blow." Darwin says—the wind "gives its ivory petals to expand." It certainly is shy of opening and only occasionally when warmed by the sun, not forced by the wind,

> It "looks up with meek, confiding eye,
> Upon the clouded smile of April face,"—

are words beautifully expressed by a poet much nearer home.

Epigœa repens, called Mayflower, and Forefather's-flower, is fast becoming well known and much used of late as a souvenir present at this season of the year, and is associated historically with the ship Mayflower of Pilgrim fame, which however we believe to be of recent application. This flower commends itself both for its delicious, spicy fragrance, as well as its beauty, and is destined to find a place in literature as well as science.

Caltha palustris, Marsh Marygold, grows on the border of brooks, has a brilliant golden cup, and first flowers about the 22d of April.

> "In that soft season when descending showers
> Call forth the *greens* and wake the rising flowers."

It is called also May-blouts, or May-blobs, and all the poetry of so fine a posy often subsides into a mess of *greens*, as it is a favorite dish with many.

Old Parkinson says "It joyeth in watery places and flowereth somewhat early." All the old botanists describe it, such as Clusius, Bauhin, Tournefort, Clayton and some with figures. Cutler imputes the yellowness of butter to the cows feeding upon it.

Aquilegia canadensis, or Columbine, was noticed by the early travelers to America, and is well described and figured by Cornuti soon after the settlement of Canada, and through him obtained its specific name; it has been from that day highly prized by the botanists and florists of Europe. Parkinson says, "it was brought out of Virginia by Master John Tradescant, and flowereth somewhat earlier than any of the garden kinds, usually by a month." There is a remarkable locality of this showy flower on the hills of the Great Pasture on the east side of the road, which is much frequented by the young during the vacation of this month, who, returning with bunches of them in their hands, remind us—

> "That spring is here, the delicate footed May,
> With its slight figures full of leaves and flowers."

Sanguinaria canadensis, Blood-root, appropriately named, as may be seen by breaking the root, which is rarely avoided in digging them up. This fine flower, as large as the Ox-eye daisy, has a deserved place in many gardens, where it gradually increases and elevates its numerous and paper-white flowers in a flat surface over the plant about four inches above the ground. Its singular root is used extensively in medicine, and probably worthily so, and the plant is often figured in medical books. Its large and deeply lobed leaves give to the plant, throughout the summer, a tropical appearance. It was carried to England and cultivated as early as 1680. Linnæus wrote to John Ellis from Upsal in 1765, "If you see Mr. Lee, ask him for the Sanguinaria, which I know is to be had in England, though I have not received it from any of my correspondents."

Violets. Two or three species of this well known genus can now be obtained from the fields. They are celebrated both by the exact botanists and the idealist. Pliny says. "There be some wild and of the field; others domestical and growing in our gardens. Garlands made of violets and set upon the head resist the heaviness of the head and withstand the overturning of the brain, upon overdrinking; yea, the very smell thereof will disperse such fumes and vapors, as would trouble and disquiet the head."

Gerard, alluding perhaps to the Pansy, then called Herb Trinitie, says they "have a prerogative above others, not only because the mind conceiveth a certain pleasure and recreation by smelling and handling them, but they bring to a liberal and gentle minde the remembrance of honestie, comelinesse and all kind of virtues." An eastern poet has said of this flower,

"It is not a flower ; it is an
Emerald bearing a purple gem."

Houstonia cœrulea, one of the most common of the spring flowers, and a universal favorite, often called Violets,—a most delicate little biennial plant, its erect and very slender stem topped off with starry white or pale blue flowers with a yellow eye, and in masses often appearing like a thin sprinkling of snow over the fields.

It does not appear to have been introduced into the Kew gardens till 1785. It is figured in Curtis's Magazine and elsewhere.

Saxifraga virginiensis. One of Parkinson's seventeen tribes of plants are the "Saxifrages, or Break-stone Plants," so called from their habit of growing in the seams or crevices of rocks, not inaptly described by Josselyn as "The New England Dayzie or Primrose, the second kind of Navelwort in Johnson upon Gerard; it flowers in May and grows amongst moss upon hilly grounds and rocks that are shady." It is an Alpine plant, this and a co-species, the S. nivalis, were among the very last flowers that greeted the eyes of Kane and his weary voyagers as they pressed onward toward the pole, beyond all vegetable life.

Erythroneum americanum, most improperly and unhappily called "Dog's Tooth Violet," a fine locality of which can be seen in the low land among bushes near Legg's Hill and the Forest River road. It belongs to the Lily tribe, and it has been suggested that it be called May Lily. It has elegant glossy leaves, blotched with purple. Josselyn, in 1672, calls it "Yellow Bastard Daffodil; it flowereth in May; the green leaves are spotted with black spots." It was cultivated in England in 1665, and is mentioned in Rea's Flora.

Feathery Catkins, from the branches of Alders, Willows, Poplars and Maples, are now for a brief period shaking their pollen to the winds, and in their graceful beauty are well worthy of study. They are occasionally mentioned with much effect in the poems of Bryant, some of whose sweetest inspirations were caught under the swaying branches of his native woods.

Rev. G. D. Wildes gave an account of a recent celebration of May-day in England.

F. W. Putnam gave a summary of a paper, presented for publication by J. A. Allen of Springfield, entitled a "Catalogue of Birds found at Springfield, Mass., with Notes on their Migrations, Habits, &c., together with a List of those Birds found in the State not yet observed at Springfield." Referred to the Publication Committee.

The proposed amendments to the Constitution were read for the second time.

Charles D. McDuffie, of Salem, was elected a resident member.

WEDNESDAY, MAY 11. Annual Meeting.

Vice President, A. C. Goodell Jr., in the chair.

Donations to the Library and Cabinets were announced.

Letters were read, from S. F. Baird, William Stimpson, J. A. Allen of Springfield and L Trouvelot of Medford, relating to the publications and the "Naturalist's Directory"; from J. E. Oliver of Lynn, N. B. Shurtleff of Boston, A. W. Dodge of Hamilton and C. F. Hartt of Cambridge, on business matters; from J. T. Rothrock of Cambridge transmitting a paper for publication in the Proceedings.

The reports of the Secretary, Treasurer, Cabinet Keeper and Curators were read and accepted.

The Secretary stated that the Society was never in a more flourishing condition than at present. The receipts from the assessments of resident members had been greater than in any preceding year, which was also the case in regard to the sales of publications. During the year thirty-seven resident, and twelve corresponding members have been elected. Six members have died, leaving the number of resident members three hundred and sixty-one. Biographical notices of the deceased members will be printed in the June number of the Historical Collections. The Secretary alluded in particular to the late venerable botanist, Dr. George Osgood of South Danvers, who had always taken an active part in the Field Meetings of the Institute, and who was extensively known as one of the Linnæan school of botanists.

Five field meetings were held during the past summer, in Swampscott, Amesbury, Salem, Newburyport, and Rockport, which were all fully attended, and acknowledged successful in the attainment of their objects. Throughout the winter months meetings were held at the Society's rooms on Monday evenings, alternating with lectures on Zoölogy from Mr. F. W. Putnam.

A course of twelve lectures was given under the auspices of the Institute, at Lyceum Hall during the last winter, as follows:—two from Prof. C. T. Jackson, on Mining; one each, from Mr. C. W. Tuttle, on Cometery Astronomy; Mr. Cleveland Abbe, on Astronomical Instruments; Capt. N. E. Atwood, on the Habits of our Native Fishes; Prof. Benjamin Pierce, on Cosmogony; Mr. Alpheus Hyatt, on the Mollusca; Mr. C. M. Tracy, on Berries; President Hill, on the Geometrical Curve; and Mr. A. E. Verrill, on Corals and Coral Reefs.

The publication of the Proceedings and Historical Collections has been continued during the year. Of the former, the first quarterly number of the fourth volume, under its new form, is ready for distribution to subscribers. The Historical Collections have now reached to number one of volume six.

The annual Horticultural Exhibition took place on the 23d, 24th and 25th of September, but owing to the great scarcity of fruit, of all kinds, the tables were not loaded as in former years, though many fine specimens were contributed, particularly of grapes, which included not only those varieties grown in the hot-house, but many choice seedlings raised by the industry and care of Edward S. Rogers, of Salem. The show of vegetables was unusually good and in great variety. Heretofore very little attention has been devoted in our exhibitions, to this class of horticultural products.

To the Library valuable additions have been made, during the year, consisting of 1603 volumes and pamphlets, received from one hundred and nineteen individuals and thirty-two societies, editors of journals, and the various departments of the State and General Government. The most valuable of the donations were, one from George A. Ward, consisting of 160 volumes in the various departments of History, and general reading: and another from the retiring Vice President of the Institute, James Upton, comprising 51 valuable volumes, principally relating to horticultural subjects.

The Treasurer presented the following statement of the financial condition, for the year ending May, 1864.

GENERAL ACCOUNT.

Debits.

Athenæum Rent, half fuel, &c. . .	$491 77
Lectures, $237 76; Publications, $699 25,	937 01
Collecting Assessments, $16 50; Gas, $8 74,	25 24
Express and Postage, $24 68; Sundries, $30 22,	54 90
To Historical Account, ...	209 78
To Natural History and Horticultural Account,	42 47
Balance in Treasury, ...	7 04
	$1768 21

Credits.

Balance of last year's account, . .	39 36
Dividends Webster Bank, $40 00; Sundries, $15 60,	55 60
G. Andrews' Legacy, $190 00; Lectures, $311 55,	501 55
Sale of Publications, ...	497 70
Assessments, ...	674 00
	$1768 21

NATURAL HISTORY AND HORTICULTURE.

Debits.

Preservatives and Taxidermy, $29 22; Cases, $44 97,	74 19
Books, $30 76; Glass, $37 48, ...	68 24
Horticultural Exhibition, ...	42 12
	$184 55

Credits.

Horticultural Exhibition, ...	90 08
Dividends Lowell Bleachery, . . .	40 00
" Portland, Saco & Portsmouth Railroad,	12 00
General Account, ...	42 47
	$184 55

HISTORICAL ACCOUNT.

Debits,

Binding, $236 08; Books, $12 00, . . .	248 08
Repairing picture frames, ...	13 50
	$261 58

Credits.

Dividends Naumkeag Bank, ...	13 00
Coupons Michigan Central Railroad, . . .	38 80
General Account, ...	209 78
	$561 58

The Cabinet Keeper reported that the specimens in the Museum were in as safe a condition, as the crowded cases of some of the departments would allow. During the year, Mr. T. M. Pond has arranged, catalogued, and labelled the North American birds, and their nests and eggs. Mr. Horace Brown has done the same with the collection of Mammalia, and had commenced to catalogue the Osteological collection when other occupations prevented his completing the work. Mr. Charles H. Higbee has arranged the Mineralogical collection, and by his efforts much has been done to increase its value. The Reptiles have been partially catalogued and named. The snakes which were sent to Professor Jan of Milan have been returned in good condition, with his identifications. Among the specimens Prof. Jan found several unknown species, descriptions of which will appear in his great work on Ophidians, in which he will give full credit to the Institute for its assistance.

During the year, an Essex County Collection has been commenced with the intention to soon have the Natural History of our county fully represented and separately arranged.

The total number of donations to the various sections of the Department of Natural History, since the last annual meeting, amounts to one hundred and twenty-six, received from eighty-six persons ; besides which, several exchanges have been made with the Museum of Comp. Zoölogy at Cambridge, and the Lyceum of Nat. History of Williams' College.

Mr. James H. Emerton, curator of Articulata, reported that the Insects had been looked over, the worthless ones discarded and the others carefully protected from injury. The large collection of Brazilian Insects had been arranged, according to their orders, in a case by themselves. The other pinned specimens have been arranged in tight boxes and drawers. The alcoholic specimens of Insects, Crustaceans, and Worms have been arranged in the central cases of the large Hall. The pinned specimens of Coleoptera, Orthoptera, and Hemiptera have been catalogued and as far as possible named. Of the Coleoptera there are 1212 species, and over 3000 specimens. Of Orthoptera, 155 species; Hemiptera, 169 species; Neuroptera, 40 species. Several hundred species of Diptera and over 2000 specimens of Lepidoptera, one half of which are from South America. The Lepidoptera have been partially named by Mr. S. H. Scudder while in the Museum of Comp. Zoölogy. The small collection of Bees has been named by Dr. A. S. Packard Jr. Several exchanges have been made with Messrs. Scudder and Packard. The collection of Spiders has been largely increased during the year, and the curator, who is specially engaged in studying this order of insects, requests contributions of specimens from all parts of the country for his work.

Mr. John Robinson, Curator of the Ethnological Department, reported that the collection under his charge had been rearranged during the year. There had been many valuable donations received from fifty-five persons. Several sub-departments have recently been commenced, and good progress thus far made in rendering them available for purposes of study and examination; the Curators request the coöperation of the members and friends of the Institution in aid of these objects, trusting that their appeal will meet with a hearty response, and that many specimens will be contributed, especially such as are evanescent in their character, and if not preserved at the time, soon disappear, and afterwards are very difficult if not impossible to obtain.

The Constitution and By-Laws as revised by the Committee were unanimously adopted.

The following officers were elected for the ensuing year:

PRESIDENT.

ASAHEL HUNTINGTON.

VICE PRESIDENTS.

Of Natural History—SAMUEL P. FOWLER. *Of History*—A. C. GOODELL JR.
Of Horticulture—J. F. ALLEN.

SECRETARY AND TREASURER.

HENRY WHEATLAND.

LIBRARIAN.

NATHANIEL J. HOLDEN.

SUPERINTENDENT OF THE MUSEUM.

F. W. PUTNAM.

FINANCE COMMITTEE.

J. C. Lee, R. S. Rogers, H. M. Brooks, G. D. Phippen, Jas. Chamberlain.

LIBRARY COMMITTEE.

J. G. Waters, Alpheus Crosby, H. J. Cross, G. A. Ward, G. D. Wildes.

PUBLICATION COMMITTEE.

A. C. Goodell Jr., G. D. Phippen, Ira J. Patch, C. M. Tracy,
Wm. P. Upham, R. S. Rantoul, F. W. Putnam.

LECTURE COMMITTEE.

A. C. Goodell Jr., Francis Peabody, G. D. Phippen, George Perkins,
James Kimball, G. W. Briggs, F. W. Putnam.

FIELD-MEETING COMMITTEE

A. W. Dodge, C. M. Tracy, S. Barden, S. P. Fowler, J. M. Ives,
G. D. Wildes, C. C. Beaman, E. N. Walton.

CURATORS OF NATURAL HISTORY.

Geology—H. F. Shepard; *Mineralogy*—C. H. Higbee;
Palæontology—H. F. King; *Botany*—C. M. Tracy;
Comparative Anatomy—Henry Wheatland; *Vertebrata*—F. W. Putnam;
Articulata—J. H. Emerton; *Mollusca*—H. F. King;
Radiata—Caleb Cooke.

CURATORS OF HISTORY.

Ethnology.

William S. Messervy, M. A. Stickney, John Robinson, J. A. Gillis.

Manuscripts.

W. P. Upham, H. M. Brooks, S. B. Buttrick, G. L. Streeter, G. D. Wildes.

Fine Arts.

Francis Peabody, J. G. Waters, G. A. Ward.

CURATORS OF HORTICULTURE.

Fruits aud Vegetables.

J. M. Ives, J. S. Cabot, R. S. Rogers, John Bertram, G. B. Loring,
S. A. Merrill, W. Maloon, A. Lackey, G. F. Brown.

Flowers.

Francis Putnam, William Mack, C. H. Norris, Benj. A. West, Geo. D. Glover.

On motion of Mr. G. A. Ward, a committee, consisting of Messrs. G. A. Ward, R. S. Rantoul, George Perkins, E. N. Walton, T. M. Stimpson, Charles Davis, G. D. Wildes, S. P. Fowler, and W. P. Upham, was appointed to present the claims of the Institute upon the public for a more liberal patronage, so that it may be better enabled to accomplish the objects of its organization.

On motion of Mr. S. P. Fowler a committee was appointed to make a collection of card photographs of the members of the Institute. Messrs. S. P. Fowler, John Robinson, and D. W. Bowdoin were placed on this committee.

Mr. S. P. Fowler was requested to prepare a paper on the Ornithologists of America, for publication in the Proceedings.

THURSDAY, JUNE 9th. Ordinary Meeting.

J. G. Waters in the chair.

The following persons, having been nominated at a previous meeting, were duly elected Resident Members: William B. Parker, James S. Kimball, Edward H. Payson, S. W. Davis, Edward L. Perkins, George P. Farrington, Mrs. John Clark, J. Ford Smith, A. G. Cornelius, Charles Roundy, James F. Hale, E. F. Roberts, J. W. Roberts, David Perkins, Jeremiah Page, Henry Morton, Benj. M. Chamberlain, A. P. Amidon, all of Salem.

Additions to the Museum and Library during April, May and June, 1864.

TO THE NATURAL HISTORY DEPARTMENT.

BROOKHOUSE Jr., ROBERT. Specimens of Glyptemys insculpta from Salem.

BURT, D. W. Male and female Attacus cecropia.

CHAMBERLAIN, JAMES. Eggs of a Mollusk and an Echinus from Beverly Bar.

CHIPMAN, R. M. Attacus luna from Salem.

Cloutman, Capt. W. R. Snake from Yangtse River, Japan.

Colcord, Mrs. H. M., of South Danvers. A Red-winged Black-bird from Danvers.

Conway, Capt. Mounted specimen of White Owl.

Cooke, C. Dried Plants, Shells and Echini from Zanzibar, Africa.

Cross, H. J. Sponge from Marblehead Beach.

Davis, Charles, of Beverly. Young Eagle taken from the nest in Beverly, June 2d.

Emerton, J. H. Insects, Helix, and 2 Salamanders from an Island off Manchester. 125 specimens, 91 species of Insects from Essex Co. 147 specimens, 32 species of Insects from Salem. A collection of Ants from Salem. 6 species of Insects and a Tree Toad from Danvers. Cyclopterus lumpus and Raia sp. from Nahant.

Glover, Geo. D. Attacus cecropia from Salem.

Hunt, Y. Attacus cecropia from Salem.

Lake, Eleazer, of Topsfield. A "Little Auk," *Phaleris microcerus* Brandt., female. Shot on Ipswich River in Topsfield, Apr. 8th.

Lendall, Ed. E., of Manchester. Coral found on Glasshead flats, Manchester.

Lyceum of Natural History of William's College. (In exchange) Skin of a Seal and several Bird's Skins from Greenland. Bird's eggs from Florida and Greenland.

Marks, Capt. T. 2 Saurians from Landana, S. W. C. Africa.

Museum of Comp. Zoology, Cambridge. (In exchange) 17 specimens, 9 species of North American Turtles.

Mason, Mrs. G. R., of Lynn. 43 specimens of dried Seaweeds and 3 of Hydroids. Native.

Nichols, H. P. 117 specimens, 40 species of Insects, and a Snake from Salem.

Northend, Miss L. H. 2 specimens Attacus cecropia from Salem.

Peabody, Alfred S., of Cape Town. Copper Ore from Cape of Good Hope. Set of pressed Ferns, Bulb of a plant, Seeds of the Blue Gum Tree, Hydroids and Barnacles from Cape Town, Africa.

Pond, T. M. A Java Sparrow.

Porter, Edw 3 Salamanders from Salem.

Putnam, F. W. A Series of Crystal Models. Lime Stone from Williamstown, Mass.

Rantoul, R. S. Portion of an Indian's jaw from West's Beach, Beverly.

Rowe, Joseph. Specimens of Storeria DeKayi from Salem.

Smithsonian Institution, Washington, D. C. A collection of 27 specimens of Building Stones and Minerals.

Stone, Frank. A Bird and 2 Reptiles from Essex Co.

Unknown Donor. 2 Loons from ?

Verrill, A. E., of Cambridge. 42 species of plants from Anticosti and Grand Menan.

Watson, F. P. Larvæ of Wasps from Salem.

Webster, Mrs. John. Lime incrustation. Seeds of Plants. Eggs of Pyrula.

West, G. W. A collection of Ants and others Insects from Salem. 929 specimens, 80 species of Insects from Salem.

Wheatland, H. Iron Ore, Lime Stone and Slag from Allentown, Pa. Fishes and Salamanders from River Clarion, Elk Co., Pa.

Wheatland, Simeon J. Attacus cecropia from Salem.

White, G. W. 85 specimens of Insects from Salem.

TO THE HISTORICAL DEPARTMENT.

Brooks, H. M. Picture of Gen. Abbot. Cartridge from Alexandria. Rebel School-book.

Brown Jr., Benj. Rebel war relics.

Cloutman. W. R. Chop Sticks and Japanese Smoking apparatus. Brick from the Porcelain Tower of Nankin.

Curwen, Henry. Ancient Button. Chinese Pipe.

Farrington Jr., G. P. Rebel Equipments.

Felt, S. Q. Native Fan from the Philipine Isles. Chinese Chop-sticks.

Goodell Jr., A. C. Coins from Half-way Rock.

Higbee, C. H. Rebel War Relics from Port Hudson.

Kilby, Wm. H., of Eastport, Me. Postage Stamps.

King, Capt. Jas. B. Silver Ornaments from the ruins of the Inca city Kuamachuco.

Mansfield, Wm. Grape and Canister Shot, Bomb, and Bullet, revolutionary relics.

Marks, Capt. Thos. A Ring of Native Copper, wrought by the Natives of the W. C. Africa.

Putnam, F. W. Calcutta Hookah.

Putnam, G. D. Breastplate of the 38th Staffordshire Reg't.

Richardson, Fred. Buttons of the Bombay Artillery.

Robinson, John. Chinese Kites.

Slocum, Eben. Ancient set of Knives and Forks in a case. Ivory Cane.

Vanvleck, H. J., of Nazareth, Pa. 7 Bills of New Jersey Continental Money, and a number of Ancient Relics relating to the early history of Nazareth, Pennsylvania.

Vent, James. Indian Arrowheads.

Williams, Israel, (Estate of) Native War Implements of the Fejee Islanders. New Zealand Chief's Blanket and Native Basket. Sword taken from the Pirates on the coast of South America.

Williams, Capt. Old Shot from Fort Sewell. Coins, Military Buttons and other Revolutionary Relics found at Ft. Pickering.

Wheatland, H. Relics from the Field of Gettysburg. 5 Catholic Medals.

TO THE LIBRARY.

ALLEY, JOHN B. (M. C.) Message and Documents, 1862–3, 4 vols 8vo; Message and Doc. Navy Department, 1862—3, 1 vol. 8vo ; Message and Doc. Dept. of State, 1863—4, 2 vols. 8vo ; do. Dept. of Interior, 1863—4, 1 vol. 8vo ; do. P. O. Dept., 1863—4, 1 vol. 8vo; U. S. Coast Survey, 1861, 1 vol. 4to ; Patent Office, Mechanical, 1860, 2 vols. 8vo ; do. Agricultural, 1861, 1 vol. 8vo ; Report of Commissioners of Agriculture, 1862, 1 vol. 8vo ; Reports on the Finances 1862—3, 2 vols. 8vo ; McClellan on the Army of the Potomac, 1 vol. 8vo ; Report on Armored Vessels, 2 vols. 8vo.

BARNARD, JAMES M , of Boston. Scilla de corporibus Marinis Lapidescentibus, 1 vol., 4to, Romæ, 1752; Griffith's Icones Plantarum Asiaticum, part IV, 1 vol. 4to, Calcutta, 1854. Griffith's Palms of British India, 1 vol. fol., Calcutta, 1850.

BOSTON, CITY OF. Boston City Documents for 1863, 2 vols. 8vo.

BROOKS, HENRY M The Drum Beat, published by the Brooklyn Fair. Feb., 1864.

BURROWS, THOS. H., Supt. of Schools, Penn. Pennsylvania School Reports for 1855, 1856, 1858, 1859, 1863, 5 vols., 8vo.

CLEVELAND, MISS M. S. Turner's North Carolina Almanac for 1864, 8vo, pamph. Also several Newspapers from Newbern, N. C.

COLBURN, JEREMIAH, of Boston. Lewis's Address N. E. Hist. Gen. Soc'y, Jan., 1864, 8vo, pamph. Albany, 1864.

COWLES, WARREN, Smethport, McKean Co., Penn. Penn. School Reports 1861, 1862, 2 vols., 8vo.

CURWEN, JAMES B. Scientific American, vol. VIII. Pamphlets 25.

DERBY, MISS CAROLINE R. Wagstaffe's Piety Promoted, 1 vol., 12mo, London, 1775 ; Phipps's The Original and Present State of Man, 1 vol., 8vo, New York, 1788. Life of Mary Neale, 1 vol., 12mo, Philadelphia, 1796. Introduction to English Grammar, 1 vol., 12mo, London, 1782. Penn's Primitive Christianity, 1 vol., 8vo, Philadelphia, 1783. Various Almanacs, Pamphlets, &c.

DOWNIE, MRS. ELIZABETH A. 16th Ann. Rep. Pittsburg Mercantile Library Association for 1864, 8vo, pamph.

FLETCHER, CHARLES. Boston Gazette for 1817, 1 vol., folio. Boston Weekly Messenger for 1813, 1 vol., folio.

GILMAN, DANIEL C., of Yale College. Several pamphlets relating to Yale College.

GOLDTHWAITE, JOSEPH A. Carpenter on the Microscope, edited by Smith 1 vol., 8vo, Phila. 1856.

GOODELL JR., ABNER C. A Manifest Book began Sept. 5, 1774, ended Aug. 2, 1775, 1 vol., folio.

HANAFORD, MRS. P. A., of Beverly. Various Pamphlets and Newspapers.

JACKSON, S. C., of Boston. 17th An. Rep. of Mass. Bd. of Education, 1 vol., 8vo, Boston, 1864.

Johnson, Mrs. Lucy P. United States Commercial and Statistical Register—several numbers.

Kilby, W. H., Eastport, Me. 7th An. Rep. of Maine Bd. of Agri. 1862, 1 vol., 8vo. New Brunswick Almanac and Register, 1 vol., 16mo. Third An. Rep. of Bd. of Agri. of New Brunswick, 8vo. Maine Legis. Register, 1864. Memorials of the Centennial Anniversary of Machias, 1 vol., 8vo.

Kimball, Miss Elizabeth. Liberator for 1863, 1 vol., folio.

Kimball, James. Systems of Building Associations Examined, 1 vol., 4to, New Haven, 1856. Fisher's Marrow of Modern Divinity, 1 vol., 16mo, Boston, 1743. Proceedings of Grand R. Arch Chapter of Mass., from Sept., 1862 to Sept. 1863, 8vo, pamph.

King, Misses Hannah and Elizabeth. Winckler, Essai sur l' Electricite, 12mo, Paris, 1748. Aldini, Essai sur le Galvanisme, 1 vol., 4to, Paris, 1804. Helvetius, Oeuvres completes, vol. I, 12mo, London, 1777.

King, Henry F. Greeley's Art and Industry of the Crystal Palace, 1 vol., 12mo. Dana's Lead Disease, 1 vol., 8vo, Lowell, 1848. Clarke on the Microscope, 1 vol., 12mo, London, 1758. Ferguson on the Microscope, 1 vol., 12mo, Edin., 1858. Wood's Common Objects of the Sea Shore, 1 vol., 12mo, London, 1859. Gosse's Evenings at the Microscope, 1 vol., 12mo, New York, 1860.

Lord, N. J. Boston Post for Feb., March, April, and May, 1864.

Nason, Wm. A., of Williams' College. Williams' Quarterly, 14 numbers. Pamphlets, 11.

Metcalf, Hiram, of Boston. The Metropolitan Catholic Almanacs, for 1851, 1853, 1854, 1855, and 1856, 5 vols., 12mo.

Moulton, Herny W. Specimens of the Blank Forms used in the Provost Marshal's office, 5th District, Mass.

Montague, Wm. L., of Amherst College. Annual and Triennial Catalogues for several years.

Packard, A. S., of Brunswick, Me. Catalogue of Bowdoin College, Spring term, 1864, 8vo, pamph.

Paine, Nath., of Worcester. 4th An. Report of Worcester Public Library, 8vo, pamphlet.

Rhode Island, Sons of, by Henry T. Drowne. Oration and Poems before the Sons of Rhode Island in New York, May 29th, 1863, 8vo, pamph.

Roberts, David. Rhee's Manual of Public Libraries, 1 vol., 8vo, Philadelphia, 1859.

Sibley, John L, of Cambridge. Report on Library of Harv. Coll., Jan., 1864, 8vo, pamph.

Slocum, Eben. Boston Patriot for 1812, 1 vol., folio.

Stone, Benj. W. Manual of New York Legis., 1864, 1 vol., 16mo. 5th and 6th An. Report, of Commissioners of Central Park, New York, 8vo pamph. An. Report of the Comptroller of New York, Jan., 1864. 4th An Report of Com. of Public Charities 8vo, pamph., New York 1864.

Souther, Henry, of Ridgway, Elk Co., Penn. Pennsylvania School Rep., 1860, 1 vol., 8vo. Annual Rail-Road Reports of Pennsylvania for 1864, 1 vol., 8vo.

Thomas, Abel C., of Hightstown, N. J. The Gospel of Slavery, 8vo, pamph., New York, 1864.

United States Treasury Department. Report on the Finances for year ending June, 1863, 1 vol., 8vo.

Ward, George A Census of the State of New York, 1855, 1 folio vol. Documentary Hist. of New York, 3 vols., roy. 8vo. Documents Col. Hist. of New York, 10 vols., folio. Colonial Records of Connecticut, 1636 to 1665, 1 vol., roy. 8vo. American Eloquence, 2 vols., roy. 8vo. Annals of San Francisco, 1 vol., roy. 8vo. Ditto of Salem, 2 vols., 12mo. Messages and Documents 30th Congress, 1 vol., 8vo. Ditto 1849, 1 vol., 8vo. Prescott's Mexico, 3 vols., roy. 8vo. Ditto Ferd. and Isabella 3 vols., roy. 8vo. Ditto Peru 2 vols., roy, 8vo. Miscellanies, 1 vol., 8vo. Wilkes' Exploring Expedition, 5 vols., roy. 8vo. N. E. Genealogical and Historical Register, 12 vols., 8vo. D'Aubigne's Reformation, 3 vols., 8vo. Morse's Universal Geography, 2 vols., 8vo. Graham's Hist. of United States 4 vols., 8vo. Wm. Ware's Julian, 2 vols., 12mo. Ditto Zenobia, 2 vols., 12mo. Dewey's Works 3 vols., 12mo. Ditto Sermons, 1 vol., 12mo. Headley's Washington and his Generals, 2 vols., 12mo. Ditto Two Wars with England, 2 vols., 12mo. Ditto. Adirondack, 1 vol., 12mo. Miscellanies, 1 vol., 12mo. Hannah Adam's Hist. of Religion, 1 vol., 12mo. Rupp's Hist. of Religion, 1 vol., 8vo. Covode's Investigation, 1860, 1 vol., 8vo. Coffin's Hist. of Newbury, 1 vol , 8vo. Brooks' Hist. of Medford, 1 vol., 8vo. Gage's Hist. of Rowley, 1 vol., 12mo. Young's Chronicles of Massachusetts, 1 vol., 8vo. Brazer's Sermons, 1 vol., 16mo. Osgood's Stud. of xn. Religion, 1 vol., 12mo. Lee's Memoirs of the Buckminsters, 1 vol., 12mo. Elder Brewster, Chief of the Pilgrims, 1 vol., 8vo. E. Watson's Men, &c. of the Revolution, 1 vol., 8vo. Hist. of Portsmouth, 1 vol., 12mo. Thom's Epistle to the Corinthians, 1 vol., 12mo. Gospel Visitant, 1 vol., 12mo. Simpson's Journey Round the World, 1 vol., 12mo. Frothingham's Sermons, 1 vol., 12mo. Verses by Rev. Jas. Flint D. D., 1 vol., 12mo. Sermons, do., 1 vol., 12mo Sermons by Rev. Thos. T. Stone, 1 vol., 12mo. Ballou on the Atonement, 1vol., 12mo. Playfair's Euclid, 1 vol., 12mo. Life of S. Judd, 1 vol., 12mo. Lay Preachers, 1 vol., 16mo. Beauties of Dr. S. Johnson, 1 vol., 12mo. Capt. Canot 20 Years a Slaver, 1 vol., 12mo. Maj. Sam. Shaw's Journal, 1 vol., 8vo. Ralph Izard's South Carolina Revolution, 1 vol., 12mo. Owen's Foot Falls, 1 vol., 16mo. Chandler's Masonic Addresses, 1 vol., 8vo. Weem's Life of Washington, 1 vol., 16mo. Girard's College, 1 vol., 12mo. Patent Office Report for 1858, 1 vol., 8vo. Morison's Life of Judge Smith. of N. H., 1 vol., 12mo. Memoirs, &c. of Rev. W. B. O. Peabody, 1 vol , 12mo. Obsequies of President Monroe, 1 vol., 12mo. Catalogue of N. Y. Soc. Library, 1 vol., 12mo. Edward's Hist. and Poetry of Finger Rings, 1 vol., 12mo.

PROCEEDINGS

OF THE

ESSEX INSTITUTE.

Vol. IV. July, August, September, 1864. No. III.

SALEM:

PUBLISHED BY THE ESSEX INSTITUTE.

B. WESTERMANN & CO., 440 BROADWAY, NEW YORK, AGENTS.

OCTOBER, 1864.

G. W. PEASE, PRINTER...ESSEX STREET.

CONTENTS.

NOTICE.

The Proceedings of the Essex Institute, containing the Records of the Meetings; the Written Communications on Natural History and Horticulture, and the Naturalists' Directory, will be published quarterly, in the present form.

The Historical Collections of the Essex Institute, are issued in Bi-Monthly Numbers, of about fifty pages each, and contain all papers relating to Historical Subjects.

TERMS.

Proceedings $2 per year. Historical Collections $2 per year.

The publications of the Institute are offered in exchange for Journals, and the publications of other Societies.

Communications and Subscriptions received by H. Wheatland, Secretary, and F. W Putnam, Superintendent, *Essex Institute, Salem, Massachusetts.*

Foreign Subscribers supplied by B. Westermann & Co.

ERRATA.

Page xxxvi, line 16, for *Phaleris microcerus* Brandt, read *Mergulus alle* Vieill.

" 83, " 7, for *Myioioctes* read *Myiodioctes*.

" 83, " 14, for *Myiodiœtes* read *Myiodioctes*.

Bellsham's Evidence of the xn. Religion, 1 vol., 12mo. Thatcher's Hist. of Plymouth, 1 vol., 16mo. Eraste 2d vol. on l' ami de la Jeaneese, 1 vol. The Lost Prince (Louis 17th) Rev. Williams 1 vol., 12mo. European Magazine for 1792, 1 vol., 8vo. Memoirs of Thos. H. Perkins, by Cary, 1 vol., 8vo. Washington's Farewell Address, (Hist. of) 1 vol., 12mo. Dr. N. L. Frothingham's Metrical Pieces, 1 vol, 12mo. Stockdale's Brit. Peerage, 1 vol., 12mo. Do. Baronetage, 1 vol., 12mo. Parson's Life of Sir. Wm. Pepperrell, 1 vol., 12mo Obsequies, &c. on Henry Clay, 1 vol., 12mo. Girard's Will Case, 1 vol., 12mo. Eliot's Biographical Dict., 1 vol., 12mo. Memoirs of General Chas. Lee, 1 vol., 16mo. John Adams' Cunningham Correspondence, 1 vol., 12mo. Brazer's Holyoke's Ethical Essays, 1 vol, 12mo. Combe on the Constitution of Man, 1 vol., 16mo. Jefferson's Notes on Virginia, 1vol., 12mo. Newhall's Letters to John Pickering on the Letters of Junius, 1 vol., 12mo. Letters to and from John Wilkes, 1762, 1 vol., 12mo. Dealings with the Dead, L. M. Sargent, 1 vol., 12mo. Emerson's Hist. of the First Church Boston, 1 vol., 12mo. Letters and Dispatches of Cortez 1 vol., 12mo. Willis's Idlewild, 1 vol., 12mo. The Landing on Cape Ann by Conant, 1 vol., 8vo. Travels in Holland, 1 vol., 12mo. Map of Holland, 1 map. Linda, (Slave) 1 vol., 12mo. Salads for the Solitary, 1 vol., 12mo. Notices on Duels by Sabine, 1 vol., 12mo. Memoirs of Lucien Bonaparte, 1 vol., 12mo. Furness' Discourses, 1 vol., 12mo. Bible News, 1 vol., 16mo. Notte's 50 Years in both Hemispheres, 1 vol., 12mo. Memoir of John W. Foster of Portsmouth formerly of Salem, 1 vol., 12mo. Life Thoughts of Henry Ward Beecher, 1 vol., 12mo. Dr. Farley's Unitarian Lectures, 1 vol., 12mo. Genealogical Register, 1 vol., 12mo. Irving's Life of Washigton, 5 vols., 8vo. Lives of American Merchants, 2 vols., 8vo. Quincy's Hist. of Harv. College, 2 vols., 8vo. Macaulay's Hist. of England, 2 vols., 8vo. Gasparin on the American Rebellion, 1 vol., 12mo.—Volumes, 161.

Waters, J. Linton, of Chicago. Adjutant General's Report, State of Illinois, 1 vol., 8vo, 1863. War Record of Illinois, to Oct. 1, 1863, 8vo, pamph. Revised Charter of Chicago, 8vo, pamph. 1863.

Wildes, J. H., of San Francisco, Cal. 11th An. Rep. of San Francisco Mercantile Library Association for 1864, 8vo, pamph.

Williams, Estate of the late Israel. Charnock's Biographia Navalis, vols. 1 to 5—5 vols., 8vo, London, 1794. Stavorinus's Voyage to the East Indies, 3 vols., 8vo, London, 1798. Whitman's Travels to Turkey, &c., 1 vol., 8vo, Philad., 1804. Turnbull's Voyage Round the World, 3 vols., 16mo, London, 1805. Naval Trade and Commerce, 2 vols., 8vo. Lathrop's Sermons, 1 vol., 8vo, Worcester, 1806. Wayland's Discourses, 1 vol., 12mo, Boston, 1833. Malham's Gazetteer, 2 vols., 8vo, Boston, 1797. Eustaphieve's Character of Peter the Great, 1 vol., 12mo, Boston, 1812. Demetrius Epick Poem, 1 vol., 12mo, Boston, 1818. Apontamentos Grammaticos e filologicos, 1 vol., 16mo, Boston, 1787. Voyage in Search of La Perouse, 2 vols., 8vo, London, 1800. Beatson's view of the War with Tippoo Sultana, 1 vol.,

4to, London, 1800. Ware's European Pilot, 1 vol., Whitehaven, 1774.

WOODBURY, EZRA J. The Works of Thomas Goodwin, 1 vol., folio, London, 1671.

BY EXCHANGE.

AMERICAN GEOGRAPHICAL AND STATISTICAL SOCIETY. Proceedings, vol. II, No 2, 8vo, pamph.

AMERICAN PHILOSOPHICAL SOCIETY. Proceedings, vol. IX, Nos. 68, 69, 70.

BUFFALO YOUNG MEN'S ASSOCIATION. 28th An. Rep., 8vo, pamph.

CANADIAN INSTITUTE. Canadian Journal for Mch. and April.

CINCINNATI MERCANTILE LIBRARY ASSOCIATION. 29th An. Report, 8vo, pamph., Cincinnati, 1864.

EDITORS. Historical Magazine, for April and May, 1864.

EDITORS. Round Table, Nos., 24 to 32.

FIRELANDS HISTORICAL SOCIETY. Firelands Pioneer, vol. 5, 8vo.

IOWA STATE HISTORICAL SOCIETY. Annals, No. 6 for April, 1864.

LONG ISLAND HISTORICAL SOCIETY. Brooklyn Manual, for 1858-9, 1859-60, 1860-1, 3 vols., 12mo. Various Documents Relating to the Brooklyn Sanitary Fair, Feb., 1864. Fifty Pamphlets.

MUSEUM OF COMPARATIVE ZOOLOGY AT CAMBRIDGE. Annual Report of the Trustees, 8vo, pamph., Boston, 1864.

PEABODY INSTITUTE. 12th An. Rep , of the Trustees, 8vo, pamph.

PHILADELPHIA ACADEMY OF NATURAL SCIENCE. Proceedings, Jan., Feb., Mch., April.

PUBLISHERS. North American Review, for April, 1864.

PUBLISHERS. Lynn Weekly Reporter. Lawrence American. South Danvers Wizard. Haverhill Gazette. Essex Banner. Toulumne Courier.

RHODE ISLAND HISTORICAL SOCIETY. Rhode Island Colonial Records vols. 1 to 7, 7 vols., 8vo. 87 Pamphlets.

SAN FRANCISCO MERCANTILE LIBRARY ASSOCIATION. 11th An. Rep., 8vo, pamph.

SMITHSONIAN INSTITUTION. Smithsonian Miscellaneous Collections, vol. IV. Smithsonian Contributions to Knowledge vol. XIII. Annual Report, for 1862.

ZOOLOGISCHEN GESELLSCHAFT, Frankfort, a. M. Der Zoologischen Garten, Nos. 7, 8, 9, 10, 11, 12 (1863,) and No. 1 (1864,)

WEDNESDAY, JULY 6. Ordinary meeting.

Francis Peabody in the chair.

The following persons having been nominated at a previous meeting by Messrs. G. A. Ward, H. Wheatland and W. P. Upham, were elected Resident Members; Daniel

Perkins, John Felt, William H. Jelly, John Chapman, Francis Boardman, Charles S. Rea, Joseph H. M. Bertram, Joseph H. Hanson, Nathaniel Brown, John H. Nichols, Edward A. Smith 2d, Francis Choate, Samuel R. Hodges, Theron Palmer, William H. Kehew, George G. Creamer, John P. Browning, Miss Elizabeth W. Treadwell, Miss Elizabeth C. Ward Jr., Thomas M. Saunders, Manuel Fenollosa, Mrs. John H. Silsbee, F. S. Peck, Jeremiah S. Perkins, Richard D. Rogers, Francis W. Tuttle, W. J. Stickney, Orrin F. Thompson, Henry Hale, Xenophon H. Shaw, John H. Downing, Willard Goldthwaite, John Francis Tuckerman, M. H. Hale, W Reith Jr., Jonathan Tucker, George C. Lord, George A. Fuller, Augustus Perry, Tristram T. Savory, E. A. Simonds, J. F. Almy, George B. Jewett, Harrison O. Flint, John P. Peabody, Henry Hubon, Henry G. Hubon, Ingalls K. Mackintire, Charles Sanders, Miss Caroline Saltonstall, James Manning, William B. Ashton, Henry K. Oliver, Nathaniel Brown Jr., Joseph F. Walden, Daniel C. Haskell, Daniel E. Clough, George P. Daniels, Charles Lamson, Peter Silver, Charles H. Glazier, George Bowker, Charles Bowker, all of Salem, and Samuel Porter of Beverly.

WEDNESDAY, JULY 13. Field meeting at East Saugus.

This meeting was attended by a company who arrived by the 10 A. M. train from Salem, Lynn, and other towns, and made their rendezvous at "Waverley Hall." From thence, dividing into parties, one pursued the route leading to "Round Hill" and the "Center," while another took a path on the easterly side of the river and visited "Pirates Glen" and other interesting spots adjacent. Others pushed their travels as far as "Dungeon Rock" in Lynn. At 3 P. M. the meeting was called to order, and Rev. S. Barden of Rockport was invited to preside during the absence of the constituted officers.

The records of the preceding meeting were read, and letters were announced from the following:

A. Agassiz, of Cambridge ; John H. Klippart, Sec'y Ohio State Bd. of Agriculture; J. Mayer & Co., of Boston; S. F. Baird of the Smithsonian Institution; Edward L. Graeff, of Brooklyn, N. Y.; E. T. Cresson, of Philadelphia; J. A. Allen, of Springfield; L. Trouvelot, of Medford; Henry White, of New Haven, Conn.; C. B. Richardson, of New York, relative to the publications. From the Maine Historical Society ; Massachusetts Historical Society ; Literary and Historical Society, of Quebec ; Lyceum of Natural History, of New York ; Corporation of Harvard College ; Trustees of Newburyport Public Library; Trustees of Boston Public Library, severally acknowledging receipt of publications. From Jere. Page ; Willard Goldthwaite ; Charles Bowker ; Francis Boardman; Geo. P. Daniels; W. J. Stickney; H. K. Oliver ; M. Fenollosa ; G. B. Jewett, accepting membership. From C. M. Tracy, of Lynn ; Mrs. P. A. Hanaford, of Reading ; S. H. Scudder, of Boston, relating to Field meetings. From Long Island Historical Society ; A. S. Packard, of Me. Hist. Soc.; Henry W. Moulton ; Smithsonian Institution, relating to exchanges and transmission of books and specimens. From J. Porter, of Wenham ; Henry A. Smith of Cleveland, Ohio ; Mrs. Hannah B. Russell ; Morris Phillips, of New York ; C. B. Preston, of Danvers ; R. M. Piper, of Nahant ; Lowell Bleachery ; Geo. A. Ward, on business matters.

Donations to the Library and Cabinets, were announced. The chair proceeded to give an account of the geology of the place, as observed by him. He spoke of the "Jasper Ledge" of "Round Hill" and its amygdaloid, and of the fine porphyries of this region. At this point Vice President A. C. Goodell Jr. arrived and took the chair.

C. M. Tracy, of Lynn, described a variety of plants and flowers gathered during the day. A fine cluster of rhododendrons being sent to the table he gave some account of the family to which it belonged and of its peculiarly fine developement among the Alps and Himalayas. Also of the other splendid members of this family, the azaleas, the kalmias, the heaths, &c.

Wilbur F. Newhall, of Saugus, gave somewhat extended remarks on the more prominent points in the history of the town. He remembered the old Newhall Tavern formerly kept here by an ancestor of his, and famous in its time, and he recollected being carried, when a little boy, to see the old building torn down. At this tavern, as he had heard the story from an ancient dame, Washington once stopped during his journey, rested awhile, and only allowed himself a cup of cold water. Mr. N. also spoke of a series of articles on the history of the town, prepared by his late father Benj. F. Newhall, and printed in the Lynn Reporter, a reprint of which is under consideration.

Joseph Dampney, of Lynn, gave some further statements in the same direction, particularly as to the first church built in Saugus, which was at the "Center."

The chair said that Saugus was a place very interesting to the antiquary, and historian. Some of the ramblers of the day had probably seen the heaps of scoria at the "Center" called the "Cinder Bank." At that spot was established the first iron foundery in the country, the scene of the labors of Joseph Jenks, one of the pioneers of American inventors. It was he who first contrived and introduced the long stiff scythe now used by mowers; and we also find record to show that he invented an "engine for the more spedye cutting of grasse," for which he sought legislative encouragement. What this "engine" was does not well appear. The foundery was a success, it would seem, and a choice relic from it is yet preserved in the family of the late Alonzo Lewis, of Lynn, to wit, the first article cast, being a small quaintly shaped iron pot.

Rev. C. C. Beaman, of Salem, gave a brief account of the delightful scenery at and about "Pirate's Glen" and also adverted to the tribe of Indians who formerly dwelt hereabout. It was said that their camps might still be traced by the imbedded clam-shells in the soil; and there were some who could recollect the last of these easy, indolent, fish-eating people, as they lingered awhile among their civilized and more powerful successors.

Rev. A. W. Bruce, of Marblehead, expressed his satisfaction at the proceedings of the day, and spoke further of the importance of preserving items of local history.

James H. Emerton, of Salem, made some statements as to the large collection of insects received by the Institute in the past year, and gave some suggestions on the preservation of specimens in this class.

P. L. Cox, of Lynn, testified to the pleasantness of the day's affairs, and paid a just and warm tribute to the memory of Benj. F. Newhall, the historian of Saugus.

Prof. John C. Holmes, of Michigan, gave some description of the tulip tree as found in that State (allusion having been made to the tree by Mr. Tracy.) He then spoke of the developement of the agricultural resources of the West, the transportation eastward of their products, and the necessity of increased facilities for this purpose.

W. P. Upham, made some remarks on the jasper and porphyry of this region.

Rev. C. C. Beaman, of Salem, called attention to the character and worth of the late Joshua Coffin Esq., the historian of Newbury, and on his motion, the Vice President of History was requested to prepare a memoir of that venerable author to be read at a future meeting.

On motion of Mr. Upham the thanks of the Institute were tendered to Messrs. Wilbur F. and Alston Newhall for their services as guides and otherwise, to the Proprietors of Waverley Hall and all our friends in Saugus for their kindness this day. Several persons were nominated for membership and the Institute then adjourned.

THURSDAY, JULY 14. Ordinary meeting.

G. A. Ward in the chair.

The following persons nominated at the Saugus meeting by Messrs. G. A. Ward, R. S. Rantoul and H. Wheatland, were duly elected Resident Members; Abraham J. Stanley, Samuel Carlen, John Mackie, C. W. Richardson, W. C. Moulton, Charles Baker, J. M. Rice, Miss Annie Treadwell, Mrs. Mary Doyle, Mrs. Chas. Hoffman, Charles Sewall, Thomas P. Newhall, Robert McCloy, James Treffren, Fred. Porter, W. D. Northend, Julian A. Fogg, J. S. Cross, Andrew H. Lord, Charles Osgood, Mrs. D. A. Neal, all of Salem; S. S. McKenzie, of Topsfield; A. W. Bruce, of Marblehead; Wilbur F. Newhall, John W. Newhall, Harmon Hall, James S. Oliver, John Westwood, and Miss Charlotte M. Hawkes, of East Saugus.

WEDNESDAY, JULY 27. Field meeting at North Beverly.

A small company of those most actively interested repaired to the neighborhood of Wenham Pond in the morning, taking the early train as far as the little village of North Beverly. These spent the forenoon in various rambles in the vicinity and being joined by a large additional force from Lynn, Salem, and other places, at about 3 P. M. the regular meeting was organized on the westerly margin of the pond under a clump of venerable pines on the grounds of Richard P. Waters Esq. Vice President, A. C. Goodell Jr., took the chair and made a few remarks, explanatory of the objects of the Institute.

After the reading of the records of the last Field meeting, and the announcement of donations to the Library and Cabinets, letters were read from the following;

Trustees of Boston Public Library, acknowledging receipt of publications; W. F. Newhall, of East Saugus; A. W. Bruce, of Marblehead; D. C. Haskell, J. F. Walden and Mrs. J. H. Silsbee, of Salem, accepting membership; R. M. Piper, of Nahant; S. Barden, of Rockport, and Wm. Lafavor, of Salem, on business matters.

Robert S. Rantoul, of Salem, read an extended essay on the History and Uses of Wenham Pond. In his remarks he spoke of the remarkable purity of its water; its permanency of level; the enormous crops of ice taken from it and the esteem in which this product is held abroad; the many notable characters associated with it in history, particularly Rev. Hugh Peters; and the singular amount of litigation that had marked the adjoining territory in the course of years.

A short but very pleasing poem by Mrs. J. H. Hanaford, late of Beverly, but now of Reading, was read by Rev. Geo. D. Wildes of Salem who prefaced it with a few remarks. The Poem was descriptive of the emotion felt by an American in Europe on meeting with a specimen ot the famous ice from these waters.

Stephen H. Phillips, of Salem, adverted to the very interesting fact that this was one of those "greate pondes," of more than ten acres in extent, whose entire freedom to all our people for fishing and fowling is guaranteed forever, first, by the "Bodye of Libertyes," drawn and promulgated by Rev. Nathaniel Ward of Ipswich in 1643, then by later enactments of the General Court, and now finally made a fixed fact by decision of the Supreme Court lately rendered. He read extracts from the manuscript opinion of the Court in the case of Inhabitants of W. Roxbury *vs.* Stoddard, bearing on this point. Thus, said Mr. P. we are in full posession of these lovely waters, for all legitimate public uses, free of cost and beyond hinderance by designing men; and this more than by all else, by the early foresight of Nathaniel Ward of Ipswich, known as the "Simple Cobbler of Agawam."

James Slade, late City Engineer of Boston, gave some interesting facts on the subject of furnishing water to cities, and said that when a tolerable source was selected, it was always found that the quantity provided by nature could be much increased by art, by the use of means to prevent loss and waste.

Rev. G. W. Skinner, of Gloucester, made some statements upon the remarkable ridge, or moraine, which runs along the shore of the lake, from near this spot to almost the northern end. He discussed its structure very fully and concluded that it was formed, during the period of drift, by the deposit of stones and gravel brought by ice-floes or field-ice, which here, restrained by the highlands, was forced to move for sometime in a kind of eddy.

C. M. Tracy, of Lynn, made some observations on the peculiar structure of the *Sarracenia* or Huntsman's Cup. He favored the idea that its pitchers, which are usually partly full of pure water, are reservoirs for the collection of dew, which may, by some natural means, be formed upon them more readily than upon other objects. The specimen before the meeting was from Cape Ann, and, despite the severe drought, had been found with its usual supply of water.

Prof. B. O. Pierce, of Beverly, had also examined the moraine spoken of by Mr. Skinner, and gave some considerations thereon, as also on the mollusca found in Wenham Pond.

Richard P. Waters, of Beverly, said this moraine had attracted the notice of Hitchcock who had pronounced it a wonderful formation; but he seemed not to have alluded to it in his writings.

Rev. C. C. Beaman, of Salem, gave some notice of the earlier proprietors of this region, and particularly of Rev. Mr. Fiske, one of the first clergymen in Wenham; also of the church records of that old parish which are still preserved.

Charles S. Osgood, of Salem, alluded to the kind entertainment given us this day, and moved the thanks of the Institute to the friends who had furnished it. The same were voted unanimously. After the nomination of several persons for membership the Institute adjourned.

THURSDAY, JULY 28. Ordinary meeting

Vice President, A. C. Goodell Jr., in the chair

The following persons nominated at the North Beverly meeting by A. C. Goodell Jr., and H. Wheatland, were elected Resident Members; Isaac Appleton, of Beverly; Geo. P. Russell, of Haverhill; Shadrach M. Cate, Ephraim Miller, James C. Stimpson and George Newcomb, of Salem. The thanks of the Institute were voted to Mr. Rantoul for the reading of his paper, on the "History of Wenham Pond" at the meeting of yesterday, and a copy was requested for publication in the Historical Collections.

WEDNESDAY, AUGUST 10. Field meeting at Gloucester.

About three hundred persons arrived in the first train from Salem, and were escorted to the Town Hall where a few remarks of welcome were made by Rev. Mr. Skinner, of Gloucester, and the divine blessing was invoked by Rev. Mr. Banvard of Worcester. The party was then dismissed for rambles and observations. Some visited the "Stage Rocks" and "Rafe's Chasm"; others rambled along the beach or in the woods in search of plants and animals.

At one o'clock the party had mostly reassembled at the Town Hall, where after appeasing the good appetites caused by the morning walks, the meeting was called to order by Rev. S. Barden, of Rockport, who made a few opening remarks.

The records of the last Field meeting were read, and donations to the Cabinets and Library announced. Letters were announced from:

New Jersey Historical Society, acknowledging the receipt of publications; J. F. Tuckerman, of Salem, accepting membership; C. M. Tracy, of Lynn; A. P. Peabody, of Cambridge, and G. W. Skinner, of Gloucester, respecting Field meetings.

G. D. Phippen, of Salem, gave a brief account of the early history of Gloucester, and then spoke of the transmutation of species among plants, holding that, while under cultivation, plants were by the hand of man, changed, so as to produce well marked varieties, yet, if left to nature's own laws, every species would remain true to the characteristics stamped upon it by the Creator, at its first appearance upon earth.

Rev. E. C. Bolles, of the Portland Nat. Hist. Society, upon being introduced, made a most eloquent, and appropriate speech, advising all to study the works of God in the field, and open their eyes to the beautiful gems at their feet. Mr. Bolles stated that he had come from Portland with his fellow member of the Nat. Hist. Society, Mr. Morse, to see how a field meeting was conducted, and hoped that his own Society would be able to follow the example of the Essex Institute.

Rev. G. W Skinner, of Gloucester, exhibited, under a microscope, some infusorial earth found on the Cape, and explained the probable origin of the deposit.

Prof. Wm. Hinks, of University College, Toronto, C. W. was introduced to the meeting, and gave an interesting, general account of the lower animals and plants, during which he stated that he was inclined, with others, to admit a fifth branch to the animal kingdom, in which the sponges and allied organisms should be placed.

Rev. Joseph Banvard, of Worcester, gave an account of the Worcester Society which had similar objects with those of the Essex Institute, and had commenced to hold field meetings. In the Worcester Society, ladies are not only admitted as members, but are elected assistant curators, and take an active part in all the meetings of the Society, reading papers, and discussing the various subjects presented. Mr. Banvard stated that he had recently seen the ants feeding upon the juices secreted by the aphides, or plant lice, and that he had noticed three distinct species of ants, each of which lived upon the secretions of a peculiar species of aphis.

Ed. S. Morse, of Portland, whose especial study is the land snails, gave an account of the collection made by himself and Mr. Bolles during the morning, stating that he had found several specimens of two very rare species of minute snails. The structure of these little snails, furnished, like most of the larger species, with a shell, which is secreted by, and is a part of the animal itself, and not a house which it can leave at will, as is commonly supposed, was explained by drawings. He also showed the position and shape of the hundreds of microscopic teeth with which the snail's tongue is furnished for the purpose of rasping its food.

Mr. Morse read the following list of Terrestrial Mollusca collected at Gloucester during the morning.

Tebennophorus dorsalis *Binney*.
Limax campestris *Binney*.
Helix striatella *Anthony*.
" labyrinthica *Say*.
" arborea *Say*.
" chersina *Say*.
" lineata *Say*.
" milium *Morse*.
Helix ferrea *Morse*.
" Binneyana *Morse*.
" exigua *Stimpson*.
Vertigo ovata *Say*.
Pupa pentodon *Say*.
Succinea Totteniana *Lea*.
" avara *Say*.
Melampus bidentatus *Say*.

A. C. Goodell Jr., called the attention of the meeting to the little neglected barnacle on the rocks, and after giving an interesting description of its structure, which he illustrated by a drawing of that portion of the animal under the shell, he favored the meeting by reading a few stanzas, *found in his pocket*, relating to the little crustacean.

F. W. Putnam, of Salem, being called upon to explain the structure of the lobster and other animals that had been collected during the day, gave a brief account of the various animals, and by a comparison of the lobster with the young barnacle, which for a short period of its life, is a free swimming animal, showed how closely related were the two, and how erroneous was the common opinion, that the barnacle was a mollusk, on account of its limy shell.

Prof. A. Crosby, of Salem, gave an account of the walk taken by his party to the rocks, where many interesting things were discovered, and several kinds of minerals collected.

Rev. S. Barden, of Rockport, exhibited a number of the minerals that had been collected, and described the structure of each.

George F. H. Markoe, of Boston, explained the various properties of the medicinal plants which he had collected, and furnished the following list of plants seen during the day.

Drosera longifolia
Drosera rotundifolia.
Leucanthemum vulgare.
Maruta cotula.
Nymphæa odorata.
Nuphar advena.
Gaultheria procumbens.
Achillea millefolium.
Asclepias incarnata var. pulchra.
Platanthera blephariglottis.
Sambucus canadensis.
Mitchella repens.
Leontodon autumnale.
Arctostaphylos uva-ursi, in fruit.
Hypericum perforatum.
Hypericum sarothra.
Elodea virginica.
Silene inflata.
Statice limonium.
Clethra alnifolia.
Cuscuta Gronovii.
Eupatorium perfoliatum.
Epilobium angustifolium.
Epilobium lineare.
Cornus canadensis, in fruit.
Scutellaria laterifolia.
Spiræa tomentosa.
Spiræa salicifolia.
Œnothera biennis.
Œnothera pumila.
Antennaria margaritacea.
Eupatorium purpureum
Impatiens fulva.
Lobelia cardinalis.
Lobelia inflata.
Lobelia spicata.
Pontederia cordata.
Sagittaria variabilis var. sagittifolia.
Vaccinium oxycoccus.
Lythrum salicaria.
Xyris bulbosa.
Solanum dulcamara.
Oxalis stricta.
Trifolium repens.
Trifolium pratense.

James H. Emerton, of Salem, exhibited a collection of about an hundred species of insects, including many species of spiders, the object of his special study, that had been collected by him during the day.

Rev. E. B. Willson, of Salem, made a few general remarks upon the usefulness of these meetings in promoting the study of Nature.

Henry W. Peabody, of Salem, was nominated for Resident Membership by A. C. Goodell Jr. and H. Wheatland.

On motion of Mr. Goodell the thanks of the Institute were voted to the Selectmen of Gloucester, for the use of the Town Hall during the day, and to Rev. G. W. Skinner and other friends in Gloucester, for kind attentions. Adjourned.

THURSDAY, AUGUST 25. Field meeting at Rockville, South Danvers.

A company of pleasant size and character gathered this day at the little chapel at "Rockville" for a series of refreshing rambles in the neighborhood of our old familiar "Ship Rock." Some of the party started for Bartholomew's Pond; others proposed to find "Wildcat Ledge" on the declivity of Prospect Hill near the line of Lynn; and some went to Spring Pond and the Aqueduct Fountains. The largest portion, probably, as generally happens, took the shortest walk, and ended their jaunt at "Ship Rock." The iron ladder and steps, provided by the Institute, are still in good order; and the shady woods around were very refreshing for a hot and dusty day.

The afternoon meeting was organized in the chapel; Rev. S. Barden, of Rockport, taking the chair. On so doing, he remarked that we had brought stones, plants and animals, and displayed them on and about the sacred desk. It might seem as if this apparent desecration needed some apology, but to him, at least, it was evident, that no antagonism existed between these elements, but most beautiful harmony. True, we seldom see it exemplified in this way. The works of God are never opposed to his word; and Nature teaches nothing in support of irreligion or vice.

The records of the last meeting were read and donations to the Library and Museum announced. Letters were announced as received from the following persons and Societies, since the last meeting:

Julian A. Fogg; John P. Browning; George P. Russell, of Haverhill; J. H. Wildes of San Francisco, accepting membership: James D. Dana, of New Haven, respecting Ordway's "Tree Protector": B. Westermann & Co. of New York; J. A. Allen, of Springfield; Raynal Dodge, of Newburyport, relating to the publications; J. D. Dana, of New Haven; James Hubbert, of Toronto; S. F. Baird, of the Smithsonian Institution; James Hall, of Albany; Vincent Barnard, of Chester Co. Pa.; Charles H. Pitman, of North Barnstead, N. H.; Wm. Dawson, of Spiceland, Ind.; Amory L. Babcock, of Sherborn; Geo. C. Huntington, of Kelley's Island, Ohio; James Lewis, of Mohawk, N. Y.; John Johnston, of Middleton, Conn.; John Haywood, of Kingston, Ohio; W. M. Beauchamp, Skaneateles, N. Y.; Wm. Muir, of Fox Creek, Mo., relating to the Naturalists' Directory: S. Jillson, of Feltonville; E. S. L. Richardson, of Chicago, Ill.; P. A. Hanaford, of Reading, on business matters.

The chair then spoke of the geology of this region; and said that he had been able to-day to verify the observation made by Messrs. Alger and Jackson in 1848, of scratches and groovings on the ledge under the eastern base of Ship Rock. These clearly proved it a bowlder; since there must have been a time when it stood elsewhere, and other materials were doing this grinding work in the place it now occupies. Under the well known rock in Gloucester called the "Whale's Jaw," similar markings are to be seen, proving the same thing. If any one doubted that such rocks had ever been transported, or that ice was an adequate agent for such work, he had only to visit Cape Cod in the winter, when in one of its harbors it might be seen at play, as it were, with a great stone, carrying it rods away and back, this way and that, with every tide.

F. W. Putnam exhibited the various animals which had been collected and explained the characters of the bream, perch and shiner, showing in what way the shiner differed from the other two, and how the perch and bream belonged to two closely allied families. He stated that the three species under consideration had a wide geographical range, only equalled by one or two other North American fishes, being found in almost all the ponds and lakes east of the Rocky mountains and south of the Arctic regions. He also made some brief statements as to the nature and habits of the several kinds of batrachians such as frogs, toads, and salamanders.

C. M. Tracy, of Lynn, made some explanation of the plants collected by the explorers, particularly of the composite family, which make ten per cent. of the world's vegetation, and were well represented to-day, by a prodigious thistle, some six feet high. A few moments were spent in considering a variety of plants reputed to cure the bite of snakes and other venomous animals. Some of them, it was stated, probably possessed a degree of virtue, while others would be but idly employed for such a purpose.

Rev. Joseph Banvard, of Worcester, said that he had seen to-day, fresh evidences of that grand principle of Nature, that all life is nourished by decay. Death and dissolution are everywhere before us. The animal dies, the plant perishes, and both are turned to mould. The rock weathers and disintegrates. Ship Rock itself is crumbling. From the dust of all decaying structures, a new order and generation of things, sentient and otherwise, springs constantly up, to fill a place and enjoy a time in the universal history. So in all things. In a sense wholly legitimate, we have lived for years on the blood and bones of our Revolutionary Fathers. To-day we are called to fertilize the soil anew with sacrificial blood, that life and enjoyment may arise for future generations. These things are often more literal than we think. When, some time ago, there was opened the grave of good old Roger Williams, the root of an apple tree was found to have travelled to the head of the coffin and penetrated all along the spine, and thence branched down the legs to the feet, being thus nourished by the material of the bones. And therefore those who ate of that tree had been unwittingly partaking of the very substance of the old Reformer. Nor in all this is there anything abhorrent to a fine and merciful sense. Nature destroys with sudden stroke, mostly, all things that can feel. She saves pain, she shows no malevolence, but only kindly transfers the life from one form to another.

Prof. A. Crosby, of Salem, gave some account of the operations of the Portland Natural History Society. This institution has excellent accommodations, and is about commencing a system of Field Meetings, much on the plan of our own. A curious feature at their rooms, is the grand table, eleven feet long by six wide, made of a single plank from the "Big Tree" of California. Prof. C. also spoke of the facilities afforded by these meetings for educational purposes, and for acquaintance with things around us which are too rarely seen in schools.

E. N. Walton, of Salem, spoke in continuation of the same subject.

The Secretary read a letter from Rev. Charles Babbage, chaplain in the army, in relation to Wenham Pond, giving some curious anecdotes of that locality, and the former residents thereabout.

On motion of C. M. Tracy of Lynn, the thanks of the Institute were voted to the Proprietors of the Rockville Chapel for the use of their premises to-day; also to the friends in the village who have favored us with their assistance and encouragement.

Henry W. Peabody of Salem, nominated at a previous meeting, was elected a resident member.

The Institute then adjourned.

FRIDAY, SEPTEMBER, 16. Field meeting at Newburyport.

This meeting had been appointed for the previous Wednesday, but postponed on account of dull weather. The company from the lower towns of the county, arriving by the morning train was quite large.

Under the efficient guidance of the Rev. G. D. Wildes, the large company were at once placed upon the route for visiting the most interesting objects in Newburyport and its neighborhood. A small party of the members whose interest was more immediately connected with the botanical and mineralogical departments, left the cars at the "Serpentine Quarry," returning thence in time for the collation and public meeting. After a general gathering at the City Hall, some of the party went on a delightful trip to Plum Island; others chose to stroll over the bridge, and enjoy the fine walk and views on the Salisbury side, and the remainder proceeded to visit the Church and Memorial Chapel of St. Paul's. The latter structure attracted special attention, from the connection with the memory of a deceased clergyman and his daughter, held in affectionate remembrance by many friends in Salem. The exquisite memorial windows of the Chapel placed as monuments to their dead, by several families of St. Paul's parish, may certainly be regarded as among the finest specimens of the stained glass to be found in this country.

From the Chapel, the party were next conducted to the beautiful grounds of the Dexter mansion, which were thrown open to them through the kindness of the proprietor, Dr. E. G. Kelley. In other particulars than this, the Institute, as on previous occasions, found themselves greatly indebted to the courtesy of Dr. Kelley. After spending some time in these grounds, the party proceeded to the Mall, the Putnam School, and thence to the beautiful Oak Hill Cemetery. None could fail to admire the new gateway, just erected through the generous gift of Mr. Tappan of New York, a native of Newburyport. None could fail to be struck with the beautiful inscription wrought in the granite entablature. We understand that the inscription was furnished by Mrs. Tappan, the daughter of the late C. W. Story Esq., of Newburyport, and we record it, as itself a testimony to a tasteful and pious culture long known to her friends:

"Until The Day Break,
And The Shadows Flee Away."

From the elevated portions of the cemetery, beautiful and extensive views of the surrounding country were obtained, embracing on the south and west the hills of West Newbury, Rowley, Ipswich, and Old Town; on the east and north the headlands of Cape Ann, the sandy shores of Plum Island, Salisbury, and Hampton; the distant Isles of Shoals, and the woods and hamlets of Salisbury, Seabrook, with the towns of Amesbury and West Newbury. After leaving the Cemetery, the Copley paintings were visited at the house of the Misses Tracy, who very kindly threw open their mansion to the large party, and furnished much valuable information as to the history of the portraits of Colonel and Mrs. Lee. Another fine portrait by Trumbull of Col. Jackson, the ancestor of the distinguished Jackson family of Boston, was seen at the same place. From this point, the route was taken to the old South Church, passing by the way the old colonial jail house in Federal street. Many of the party visited the tomb of Whitfield, where the remains of the great preacher, together with those of Prince and Parsons, were seen. After testing the quality of the whispering gallery in the church, the party proceeded to the old Tracy Mansion, once honored by the presence of Washington, Talleyrand, Chateaubriand, Louis Philippe, LaFayette and others.

This venerable mansion, now occupied, in part by the Rev. Mr. Fletcher, the distinguished traveller in Brazil, is soon to be used for the purposes of the Public Library; alterations to that effect being now made. We hope to see in connection with the valuable Public Library of Newburyport, a flourishing branch of the Essex Institute.

After viewing other localities of interest as connected with the literary, professional and commercial history of the city, the party returned to the City Hall, where the large hospitality of their friends in Newburyport had made excellent provisions for a noonday repast.

The afternoon meeting was called to order in the City Hall, about 2 1-2 o'clock, and Rev. George D. Wildes, of Salem, was invited to occupy the Chair. On assuming that place he made some remarks in explanation of the plan and practice of the Institute and the influence exerted by its meetings on the community around.

Donations since the last meeting were announced and letters were read from the following:

A. S. Packard Jr., of Brunswick, Me.; G. C. Huntington, of Kelley's Island, Ohio; J. D Dana, of New Haven; J. A. Allen, of Springfield; Thomas Barlow, of Canostota, N. Y., in relation to the publications; Smithsonian Institution, acknowledging the receipt of publications; Lyceum of Natural History of New York; S. Barden, of Rockport; W. H. Prince, of Northampton; John L. Russell; Mrs. E. H. Derby, of Auburndale, on general business; A. L. Babcock, of Sherborn; Thos. Gile, of Washington; Hiram A. Cutting, of Lunenburg, Vt., on exchanges of books and specimens.

F. W. Putnam, explained the structure of the galls found on the leaves and stems of plants, and the habits of the gall flies. He also spoke of the habits of the Aphis, Coccus and other insects injurious to vegetation.

Rev. S. Barden, of Rockport, had been to the "Devil's Den." But there was nothing there infernal; it was a place of unmixed beauty. He was glad to see the clergymen of this place interested in the pursuits of this day;

they have saved Newburyport to the cause of science. While laboring with his hammer at the ledge he had been cheered by the presence of some of them, and encouraged to open more fully the wealth of that spot. There were beautiful specimens of serpentine, as well as asbestos, or amianthus of a fine description. He exhibited an elegant vase made from the serpentine by Mr. Osgood, of Newburyport, and pronounced it equal to anything of the kind to be seen elsewhere.

Dr. H. C. Perkins, of Newburyport, said that every boy in the place had at some time been to the "Devil's Den," which few here know as a serpentine quarry. It was opened for lime exclusively and worked for some time. It furnished besides serpentine and asbestos, some very good steatite and dolomite. The celebrated Jacob Perkins, once of Newburyport, made paper from this asbestos and printed some bank-notes on it which were incombustible and served to surprise his friends.

Rev. Artemas D. Mussey, of Newburyport, expressed his deep satisfaction in the meeting and its purposes. He could not doubt its effect on those who attended, especially on the young ; and he hoped a branch society, or something like it might be formed and sustained in this place.

Rev. J. S. Spalding, of Newburyport, had fortunately met the party at the "Den" and highly enjoyed the enthusiastic activity of those who composed it. If all the members of the Institute were equally engaged and successful, the best results must follow. There are young men in Newburyport engaged in science and natural history. They have made fine collections of birds' eggs including many rare kinds and if directed and encouraged by some systematic society, they would do much for themselves and the cause of knowledge.

Rev. C. C. Beaman, of Salem, thought the Essex Institute could not fail to be greatly cheered by such language as that of the Newburyport people to-day. The historical side of our society well deserves encouragement. We are at work to preserve a worthy past by gathering and securing every relic of historic value.

Rev. Mr. Spalding, said that Essex North was rich in archeological wealth. Its history was both valuable and available. Felt, in his annals, had made some statements as to John Barnard, a celebrated teacher of the early times; but recent researches have corrected him in this matter and identified parties very differently.

Rev. John N. Sykes, of Newburyport, was glad to see the activity of the young men who took part in the operations of the Institute. The benefit of such employment in youth must be great. They would form habits of observation, which in after life would be of the greatest advantage.

C. M. Tracy, of Lynn, gave some explanation of the plants gathered by the explorers, alluding in particular to the asters, goldenrods and other autumnal flowers, and discussing somewhat the relations of the oaks and hickories. He also spoke of his visit to the garden of Dr. E. G. Kelley, in which were noticed, among the many interesting objects there found, the beautiful and finely grown hedges of hemlock, spruce and other evergreens also one of weigelia, this last in the time of flowering must have presented a splendid appearance.

Dr. Perkins said every one ought to study Natural History. It was the greatest source of comfort amid pain, sorrow and affliction, that he had ever known. When the botanical specimens were just now brought forward, they seemed to him like old friends. He remembered that forty years ago, he left Cambridge with a classmate and botanized from thence to Newburyport, losing the way in the ardor of the pursuit.

The Chair added some further thoughts on the Institute as a means of education. Such an institution forms the best of safeguards for the young and developing minds. The love of science will live every where. He had seen, in the icy fastnesses of the Alps, the little band of German students, on their vacation from the Universities, camping in the mountain valleys and enjoying their explorations with a zest that made him almost envious. Yet this enjoyment is not all, for modern science is not pleasurable only; it is eminently practical and therefore eminently useful. Encourage its growth among the people and you give them at once both happiness and power.

Stephen B. Ives, of Salem, offered the following resolutions, which were unanimously adopted.

Resolved, That the sincere thanks of the Essex Institute be presented to the City Council, of Newburyport, for the use of the City Hall, for its meeting here this day.

Resolved, That the most grateful acknowledgements of the Institute be presented to those kind friends in Newburyport, whose attentions in making the most ample, and tasteful arrangements for the field meeting, and, in providing bountiful and elegant refreshments, have rendered the present meeting among the foremost in interest and encouragement in the history of the Society.

Resolved, That the thanks of the Institute are especially offered to Mrs. D. T. Granger, Mrs. Pearson, Mrs. Nourse, Mrs. W. Horton, the Misses Tracy, of the ladies; and to the Messrs. G. J. F. Colby, E. S. Moseley, E. G. Kelley, D. T. Granger, Charles Wills, C. H. Bailey, J. H. Frothingham, J. Bogardus, J. Horton, and others who have so largely contributed to the gratification of the Institute in its present meeting.

After the nomination of members the meeting adjourned.

WEDNESDAY, SEPTEMBER 21. Ordinary meeting

Joseph G. Waters, in the chair.

William Whitaker, Thomas L. Perkins and William H.

Emmerton, of Salem, and John S. Allanson, of Marblehead, nominated at a previous meeting were duly elected Resident Members.

FRIDAY, SEPTEMBER 30. Special meeting.

The President, A. Huntington, in the chair.

The president stated that the object of our assembling this evening was to take some suitable notice of the recent sudden decease of our late associate member GEORGE ATKINSON WARD, of Salem. Mr Ward was one of the original members and very active in the organization of the Essex Historical Society. He removed to New York in 1823 to engage in business in that metropolis. He returned to Salem, in November last to spend the remainder of his life among the scenes and friends of his youth; since that time he has renewed his interest in the doings of the Institute and by his zeal and industry has largely contributed to its success.

Rev. George W. Briggs moved that a committee be appointed to prepare resolutions and a memoir to be presented at some future meetings, accompanying the same with appropriate remarks.

Francis Peabody, in seconding the motion, alluded principally to Mr. Ward's previous residence in Salem, his interest in the Institute and in all measures conducive to the intellectual and moral culture of his native place.

Rev. George D. Wildes stated that his acquaintance with Mr Ward was recent, but during that time he had seen much of him both in his walks and in visits to his home, and bore testimony to his worth and character as a citizen and a friend.

A. C. Goodell Jr. followed in remarks of a similar import and suggested that the committee consider the propriety of providing a portrait of Mr. Ward to be placed in the rooms of the Institute.

The motion of Mr. Briggs, seconded by Mr. Peabody and amended by the suggestion of Mr. Goodell, was unanimously adopted, and Messrs. C. W. Upham, A. Huntington, A. C. Goodell Jr., G. W. Briggs and Francis Peabody were appointed on said committee.

On motion of Mr. F. Peabody, Mr. C. W. Upham was appointed, in place of Mr. G. A. Ward deceased, on the committee to which was referred the "consideration of the authenticity of the tradition that the frame of the old Building in rear of Boston street is that of the first meeting house in Salem."

The committee on resolutions was authorized to call meetings whenever it may be prepared to report. Adjourned.

Additions to the Museum and Library during July, August and September, 1864.

TO THE NATURAL HISTORY DEPARTMENT.

Allen, J. A., of Springfield. 32 specimens, 9 species Reptiles from Springfield. 1 specimen Trout, young.

Babcock, Amory L., of Sherborn. (In exchange) Several fresh water Shells. Specimens of Gryllotalpa borealis and other Insects and Spiders, 3 Jumping Mice, Embryos of Native Birds from Sherborn, Mass. Body of Little Ant-eater and several Nuts from Surinam. Fossil Coral from Kansas.

Barden, Rev. Stillman, of Rockport. Specimens of Pyrhoclose, Smoky Quartz, Pyrites, Fluorspar, &c., from Rockport.

Bolles, Rev. Edwin C., of Portland, Me. 8 Specimens Helix hortensis from Broom Corn Island, Casco Bay. 3 valves of Pecten icelandicus, 4 do. of Mytilus edulis, 3 do. of Saxicava distorta *Say*, 3 do. of Astarte laurentiana *Lyell*, from the Post Pliocene, Canal St., Portland, Me. 3 Specimens of Macoma fusca, 4 do. Muscula antiqua *Mighels*, 11 do. Leda portlandica *Hitchcock*, from the Post Pliocene, Land Slide, Westbrook, Me.

Bowditch, Mrs. Rebecca. Specimen of Limax from Salem.

Briggs, Mrs. Adaline, of S. Danvers. 2 Specimens Attacus cecropia.

Brown, Benj. Fossil coral.

Brown, Horace. Specimen A. cecropia.

Byrnes, Clifford C. 2 specimens Slag. Iron found among coal.

Carlen, Samuel. Brown Bat taken in Salem.

CHIPMAN, R. MANNING. Flowers of Linnæa borealis from Westford, Mass.

CREAMER, MRS. F. M. Cones and twigs from the "Great Pine of California," also a string giving the exact circumferance of the tree from which they were taken.

DERBY, MRS. M. A., of Auburndale. Deer's horns from Minnesota. Specimen of coral.

EMERTON, JAMES H. 112 specimens, 44 species Insects, collected at the field meeting in East Saugus, June 13. 23 specimens Insects, 23 specimens 2 species Ants, 1 larva of Cicindela from Salem. 58 specimens, 34 species Insects collected in Beverly. 132 specimens, 74 species Insects collected at the Gloucester field meeting.

EMMERTON, W. H. Specimen of Walking-stick, Bacunculus femoratus, from Salem.

FARRINGTON, MISS A. W. B. Specimen of Attacus cecropia from Salem.

FLINT, G. F. Specimens of Eudryas grata.

FROST, MRS. L. A. Clay from Talahama, Tenn.

GOODELL JR., A. C. Nest of Wasps from Ipswich.

GRANT, HENRY. Fossil Mollusks from Lake Champlain.

HALL, CAPT. W. H. 6 Starfishes and Embryo Whale from West Coast of Africa.

HAMMOND, CAPT. JOSEPH. Fishes, Crabs, Starfishes and Mollusks from Baker's Isle, South Pacific. Flying-fish, North Atlantic. Several Fishes, Crustaceans, &c., from off the coast.

HANAFORD, MRS. P. A. Specimen of Chauliodes pectinicornis.

HASKELL, JOSHUA, of Marblehead. 5 specimens of Insects collected at the field meeting at Wenham Lake.

HIGBEE, CHARLES H. 3 specimens of Solitary Bees and specimen of Attacus Promethea from Salem.

KIMBALL, MRS. ENOCH F., of Wenham. Nest of Chimney Swallow.

KING, H. F. 4 specimens, 2 species Coleoptera from Gorham, N. H.

LAKE, CHARLES H , U. S. V. Specimens of Galena, Blende, Pyrites, Mica, Limestone, Tourmaline, Hematite and Fossils from the vicinity of Little Rock, Arkansas.

LEAVITT, MRS. Larva of Cerura borealis from Lexington.

LEE, JOHN C. Humming Bird from Worcester.

LEFAVOR, JOSEPH. Specimen of Cicada pruinosa.

LEWIS, I. P. Large Pearl from a Quahaug.

LORD, GEORGE R. Specimen of Monohammus sp.

LOWD, MARK. Nest and specimens of Hornets.

MERCHANT, ADDISON, of Gloucester. Barnacles and Shells from Banks of Newfoundland.

NICHOLS, H. P. 297 specimens, 148 species Insects, 2 malformed Hen's eggs, 40 specimens 3 species Salamanders, 20 specimens Fish, collected in Bethel, Vt. 73 specimens, 40 species Insects from Salem.

OSGOOD, J. C. Nest and eggs of a Wren from Salem.

PARKER, CHAS. Specimen of Walking-stick, living female.

PEASE, W. H., of Honolulu, Sandwich Isls. 29 species of Land Shells from Tahiti. 71 species of Marine Shells from the Pacific Islands. Several specimens of each species, all named and several types of new species.

PERKINS, HENRY W. Full grown larvæ of Attacus cecropia.

PUTNAM, F. W. 2 specimens of a large Aphis, with eggs and cast off skin from Salem. Quartz, Pyrites with gold, from Rangely, Me. 49 specimens of Spiders from the northern parts of Maine.

PUTNAM, CAPT. W. H. A. Collection of over 500 specimens of Coral and several Shells from Singapore, E. I. 2 specimens Forficula. Several hundred Crustacea and several Fishes from soundings off the coast.

ROBINSON, ASA P., Specimen of Nepa from Grafton Lake, Me.

ROBINSON, JOHN. 73 specimens, 50 species Insects from Salem.

RUSSELL, JOHN W. Full grown larvæ of Attacus cecropia.

SAFFORD, JOSHUA. Coal with vein of Sulphuret of Iron.

SAVAGE, MISS. Specimen of Walking-stick, female with eggs, from Salem.

SILSBEE, WILLIAM. Nest of Hornets with about 1000 specimens in different states of growth.

SMITH, HENRY. Specimen of Prionus laticollis.

SMITH, LAWRENCE P. Specimen of Attacus cecropia.

STICKNEY, M. A. Specimens of Pterogorgia and Plexura from the Cape Verd Islands.

STONE, FRANK. 85 specimens, 34 species Insects from Salem. Specimen of young Turtle from North Reading.

STONE, DR. LINCOLN R., U. S. A., Gallipolis, Ohio. Specimen of Sphinx quinquemaculata from Gallipolis.

SYMONDS, S. S. Specimens of Pelecinus sp. and Philampelus satellitia from Salem.

TRACY, C. M., of Lynn. Specimen of Scolopendra sp.

TRUE, JOSEPH. 13 specimens, 4 species Hymenoptera from Salem.

WATSON, FRANK. Specimen of Monohammus sp. from Salem.

WHITE, GEO. M. 60 specimens of a Beetle from Milkweed, Salem.

WILSON, MISS ALICE. Specimen of Cicada pruinosa from Salem.

TO THE HISTORICAL DEPARTMENT.

CHAMBERLAIN, JAMES. 2 Postage stamps, Cape of Good Hope and Victoria.

CHIPMAN, R. M. Grains of Corn from the grave of an Indian supposed to have been buried 400 years.

CREAMER, GEO. G. Piece of the Stone steps down which Gen. Putnam rode when pursued by the British during the Revolution, Greenwich, Conn.

FELT, S. Q. Piece of Palmetto wood from the Rebel ram Merrimac.

HAMMOND, CAPT. JOSEPH. Model of Canoe and native Spear Sandwich Is.

PUTNAM, PERLEY, (Estate of) 3 Weapons from the Feejee Islands.

PUTNAM W. H. A. 10 cent Postage stamp of Netherlands India.

RANTOUL R. S. Netherland Copper Coins.

WATERS, R. PALMER, of N. Beverly. Helmet of a British soldier.

WILLIAMS, ———. 4 shot, 2 fragments of shot and 1 fragment of Cannon from the old Ft. Pickering, Salem.

TO THE LIBRARY.

ADAMS, SAMPSON & Co, of Boston. N. Y. State Business Directory, 1 vol. 8vo, New York, 1859. Fall River Directory, 1864, 1 vol. 16mo. Taunton Directory, 1864, 1 vol. 16mo. Lawrence Directory, 1864, 1 vol. 16mo. Manchester Directory, 1864, 1 vol. 16mo. Charlestown Directory, 1864, 1 vol. 16mo.

CLOUTMAN, WM. R. Hoffman's Shopping Dialogues in Japanese, Dutch and English, 4to. London, 1861. Van Reed's collection of Phrases in English and Japanese, 1 vol. 8vo.

DROWNE, CHARLES, of Troy, N. Y. Annual Register of the Rensselaer Institute, 1864, 8vo, pamph.

FOOTE, CALEB. Files of the County Papers for several months.

GIBBS. J. W., of New Haven. Family Notices by W. Gibbs of Lexington, 8vo, pamph. 1845.

HANAFORD, P. A., of Reading. Bible Society Record, nine numbers. Dwight's Open Converts, 1 vol. 16mo, New York, 1836. Stone, W. L., Matthias and his impostures, 1 vol. 16mo, New York, 1835. 22 Pamphlets, also several Newspapers.

HOLDEN, N. J. Proceedings of Am. Anti-Slavery Society at its 3d decade. 8vo, pamph. New York, 1864.

HOLMES, JOHN C. 2d Annual Rep. of Secretary of Michigan State Board of Agriculture, 1 vol. 8vo. Lansing, 1863. Boston Daily Evening Traveller, for 1850, 2 vols. folio.

HOLMES, THOMAS, (Estate of) Historie de France par Anquetil, 15 vols. 16mo, Paris, 1822. Memoires pour Servir a l' histoire de France sous Napoleon, Tom 1—6 ; Tom 2, notes—7 vols. 8vo, Paris, 1823. Gourgaud's examen critique de l' ouvrage de Segur, 1 vol. 8vo, Paris, 1825. Bonnycastle's Algebra, 1 vol. 12mo, Phil., 1806. Letellier Grammaire Francoise, 1 vol. 16mo, Tournay, 1816. Gilleland's Counting House Assistant, 1 vol. 12mo, Pittsburg, 1815. Spanish Grammar by Jos. Giraldel Pino, 1 vol. 12mo, Phil., 1795. Veneroni's Complete Italian Master, 1 vol. 12mo, London, 1791. Bonnefoux, Seances Nautiques, 1 vol. 8vo, Paris, 1827. Several Log Books. Pamphlets, &c.

KLIPPART, J. H., Cor. Sect'y Ohio State Bd. of Agric. Ohio Agricultural Reports for 1853, 1856, 1857, 1858, 1859, 1860, 1861, 1862, 8 vols. 8vo.

MANN, MISS ELIZABETH N. Andover Advertiser, from 1857 to 1863 incl. 7 vols. folio.

MANNING, R. C. Cooper's Surgical Dictionary, 2 vols. 8vo, New York,

1832. Ballou's Candid Review, 1 vol. 12mo. Orton's Discourses, 1 vol. 12mo, Boston, 1816. 14 Pamphlets.

MORSE, EDWARD S., of Gorham, Me. Observations on the Terrestrial Pulmonifera of Maine, by E. S Morse, 8vo, pamph., Portland, 1864.

MUDGE, B. F., of Quindaro, Kansas. 1st Cat. of Officers and Students of Kansas State Agric. College 8vo, pamph. 3d Ann. Rep. of Sup't. of Public Instruction of Kansas, 8vo, pamph. Cat. of Baker's University. The Rocks of Kansas, by Swallow and Haven, 8vo, pamph., St. Louis, 1858.

MUNSELL, JOEL, of Albany. Catalogue of Library of Philom. Soc. of Union College, 1863, pamph. Baker's Address to Chem. Soc of Union Coll. July, 1863. Annual Catalogue of Columbian Coll. 1862—3. Albany Female Academy Report of Exam. June, 1863. Twenty-five pamphlets.

NASON, WILLIAM A., of Chicago, Ill. The Gulielmensian No. 8, May, 1864. The William's College Quarterly for June, 1864.

PACKARD, A. S., of Brunswick, Me. Catalogus Collegii Bowdoinensis, MDCCCLXIV, 8vo, pamph.

PACKARD JR, A. S., of Brunswick, Me. Synopsis of the Bombycidæ of U. S. A., by A. S. Packard Jr. 8vo, pamph.

PARSONS, G. W. "The Cartridge Box," printed at U. S. Army Hospital, York, Pa, 1864, several numbers.

PHILLIPS, STEPHEN H. Proceedings of National Union Convention at Baltimore, June, 1864.

PUTNAM, ELBRIDGE. The old Franklin Almanac for 1860—64 inclusive, 8vo, pamph.

PUTNAM, MRS. EBEN. Several Pamphlets.

PUTNAM, PERLEY, (Estate of) Nouvel Abecedaire, 1 vol. 12mo, Phil., 1811. Reuss on the trade between Great Britain and U. S. A., 1 vol. 8vo, London, 1833. Duane's Infantry Regulations, 1 vol. 8vo, Phil., 1813. Life of Moreau, 1 vol. 12mo, New York, 1806. Rawson's Military Duty, 1 vol. 8vo, Dover, 1793. Hawney's Measurer, 1 vol. 12mo, Baltimore, 1813. Steuben's Regulations 1 vol. 12mo, Boston, 1802. Vose's Astronomy, 1 vol. 8vo, Concord, 1827. Fisher's Military Tactics, 1 vol. 8vo, New York, 1805. Gray on the Revelations 1 vol. 12mo, Newburgh, 1818. Trial of Gen. St. Clair, Aug. 25, 1778, 1 vol. fol. Phil., 1778. History of Revolution in France, 1 vol. 8vo, Boston, 1794. 79 Pamphlets.

SLOCUM, EBEN. Cooper's Naval History, 2 vols. 8vo, Phil., 1840. Ditto continued to 1853, 1 vol. 8vo, New York ,1853. Browne's Whaling Cruise, 1 vol. 8vo, New York, 1846. Frost's Naval Biography, 1 vol. 8vo, Phil., 1844.

STONE, BENJ. W. Philadelphia Directory, for 1848, 1855, 1859, 1860, 1861, 1862, 6 vols. 8vo.

SWETT, JOHN, of San Francisco. 1st An. Rep. of Sup't. of Pub. Instruction of California, 8vo, pamph. Sacramento, 1863.

SYMONDS, EDWARD. Several Almanacs.

Tittle, Miss S. J., of Beverly. 12 Pamphlets.

Trask, Amos. Moore's Navigation Improved, 8vo, pamph. Salem, 1815.

Tucker, Col. James T., U. S. Volunteers. Journal and Proceedings of a Convention for a Revision of the Constitution of Louisiana, 8vo, pamph. New Orleans, 1864.

Tucker, Jonathan. Opening Address by the President at Illinois State Agric. Soc Fair at Decatur, 1864, 8vo pamph.

Tucker, William P., of Portland, Me. Catalogus Collegii Bowdoinensis, MDCCCLXIV, 8vo, pamph.

United States, Department of State. Diplomatic Correspondence, 1863, 2 vols. 8vo, Washington, 1864.

Wade, Misses, of Ipswich. Frisbie's Oration on Restoration of Peace, in 1783, 8vo, pamph. Dana 's Eulogy on Washington, 1800, 8vo, pamph.

Ward, Charles. Journal of Commerce Jr. for several months. Essex Statesman vol. 1 fol. Salem, 1863—4.

Ward, George A. The Giles Memorial by John A. Vinton, 1 vol. 8vo, Boston, 1864.

Waters, J. Linton, of Chicago. Annual Statement of Receipts and Expend. of Chicago, from Apr. 1, 1863 to Apr. 1 1864, 8vo, pamph. Catalogue of Library of Chicago Young Men's Association, 8vo pamph. Chicago, 1856, ditto 1859. Chicago Revised Charter, 8vo, pamph. 1863. 30 Pamphlets.

Wightman, W. J., of Reading. 11 School and other Reports of Reading.

BY EXCHANGE

American Antiquarian Society. Lincoln's Address on C. C Baldwin, 8vo, pamph. Jenks' Address Oct. 23, 1813, 8vo, pamph. Proceedings at Meeting April 1, 1864, 8vo, pamph.

American Philosophical Society. Proceedings, vol. ix, No. 71.

Canadian Institute. The Canadian Journal for July, 1864,

Editors. Historical Magazine, for July, Aug., and Sept, 1864.

Iowa State Historical Society. The Annals of Iowa for July, 1864, 8vo, pamph.

Long Island Historical Society. 1st An. Rep. of Directors, Librarian, &c., May, 1864, 8vo, pamph.

Montreal Society of Natural History. Canadian Naturalist and Geologist, for Feb., Apr , June and Aug., 1864.

New Jersey Historical Society. Proceedings vol. ix, No. 6, 8vo, pamph.

New York Lyceum of Natural History. Annals vol. vii, Nos. 13—16. Vol. viii, No. 1.

New York Mercantile Library Association. 43d Annual Report, July, 1864, 8vo, pamph.

Philadelphia Academy of Natural Science. Proceedings for May, June, l August, 1864.

PORTLAND SOCIETY OF NATURAL HISTORY. Proceedings, vol. 1, pages 97 to 128 incl.

PUBLISHERS. North American Review for July, 1864.

QUEBEC LITERARY AND PHILOSOPHICAL SOCIETY. Transactions, New Series, vol. I, Nos. 1 and 2.

WILMINGTON (DEL.) INSTITUTE. Annual Report April, 1864, 8vo, pamph.

MONDAY, OCTOBER 10. Evening meeting.

The President, A. Huntington, in the chair.

Records of previous meeting read and donations to the Museum and Library were announced.

Letters were read from the following:

Chas. H. Lake, of Little Rock, Arkansas; J. A. Allen, of Springfield; W. H. Dall, of Marquette Co. Mich., relating to donations of specimens: J. H. Hickcox, of the New York State Library, Albany; S. J. Young, Librarian of Bowdoin College; Secretary of the American Philosophical Society, Philadelphia, relating to exchanges of publications: Prof. S. F. Baird; J. H. Thompson, of New Bedford; Sam'l Clarke, of Milwaukee, Wis., relating to the Naturalists' Directory: Miss Lucy Treadwell, of Salem; Miss A. L Coffin, of Newbury; J. E. Oliver, of Lynn; J. W. Young, of Worcester; S. Tenney, of Cambridge; James Lewis, of Mohawk, N. Y.; Rev. E. C. Bolles, of Portland, on business matters: A. L. Babcock, of Sherborn; Dr. A. S. Packard Jr., of Brunswick, Me.; Theo. Gill, of Washington; W. Hoxie, of Newburyport, relating to exchange of specimens.

Albert B. Russell, and Miss Lucy Treadwell, of Salem, and Theodore Attwill, of Lynn, having been nominated at a previous meeting were elected Resident Members.

Mr. Putnam communicated a paper from Mr. Alpheus Hyatt Jr., entitled "Remarks on the Polyzoa of New England" In this paper, which was referred to the committee on publication, Mr. Hyatt describes and figures several new species of *Cristatella* and *Plumatella* from Cambridge, Mass., and Norway, Me. For these species he proposes the names of *C. ophidioidea*, *P. hyalina* and *P. pennissewasseensis*. Mr. Hyatt also describes the anatomy of the genera *Cristatella* and *Pectinatella* and discusses their relations, as naked Polyzoa, to the remaining genera of the sub-order Lophopea.

It was voted that meetings be held on the second and fourth Monday evenings of each month until otherwise ordered, and that all persons interested be invited to attend.

The President, from a committee appointed at the last meeting, reported that the Hon. C. W. UPHAM had consented to prepare a memoir of Mr. Ward, and was desirous of receiving any contribution that would aid in its preparation. After a few additional remarks, in which he stated that Mr. Ward was born at Salem, March 29, 1793, and died at Salem on Thursday evening, September 22, 1864, he submitted the following resolutions:

Resolved, that the members of this Institute received with deep and unaffected sorrow intelligence of the recent and very sudden death of our friend and associate, GEORGE ATKINSON WARD; and desire, by these proceedings, to express our high appreciation of his character and worth as a man and citizen, and our very great respect for his memory. As one of the original and prominent founders of the Essex Historical Society, in whose behalf he early enlisted with all his accustomed energy and enthusiasm, and to whose interests he was strongly committed, and as the last survivor of the founders of that institution, since merged in our body, it is especially fit and becoming, that we who have thus entered into these his early labors, should mark, with suitable testimonials of regard and respect, the event of his death, so sudden and startling to his friends and to this community, and so much deplored by us all. Descended from one of the most ancient and honored families of Salem, he was always ready and prepared, by his accurate and full knowledge of her annals from the earliest days of the Colony, to vindicate her character and good name; and whether at home or abroad, he was ever steadfast to the traditions, memories, and principles of the place of his birth. Endowed with the most genial qualities, with high executive ability, and with large practical and business capacities, he early sought a fitting sphere for their exercise and developement in the commercial metropolis of the country; and after walking in the high places of commercial life for more than thirty years, with varying fortunes and success, but always with honor and integrity, never too busy to foster and cultivate the studies aud tastes of his earlier life, or to engage in those works, which in all communities are required and expected at the hands of men of public spirit, and enlarged views, he came back here, but little less than one year ago, to a new generation—to die in his native and beloved town, and to be here gathered to his kindred and fathers. Although suffering from disease and infirmity he was still the same genial companionable and enthusiastic man as ever, in all good words and works, and betook himself at once, with all the zeal of his youth, to the care, culture and growth of this child of his earlier days, as one of the departments and functions of the Institute. How he labored to extend its means and usefullness, and to enlarge its boundaries; and how he commended it to the regards, support and encouragement of our people we are all this day his witnesses. He had performed the same work on a larger scale, many years ago, for the Historical Society of New York, by presenting with great attractiveness, and in his fervent and glowing manner, its objects and labors to the culture and wealth of that city, thus greatly augmenting its means, and largely aiding it in entering on that career of usefulness and renown for which it has since been so much distinguished. The hand of our friend and associate was strongly in that earlier work of revival and reconstruction; and it was only in renewal of similar labors, years before, in the formation of the Essex Historical Society. It is an affecting incident, that his very last days and thoughts were employed in preparing illustrative memorials of the first meeting house of the First Church in Salem (and the first Congregational Church founded on the Western Continent,) the frame of which is now being reërected and covered for preservation on the grounds of the Salem Athenæum, in the rear of Plummer Hall, under the direction of a committee of the Institute, a work which he had undertaken, as a labor of love, and in which he was engaged at the very moment of the fatal attack.

Resolved, That a man of a character so strongly marked as that of our deceased friend, and who has so impressed himself in various ways and degrees of usefulness on his day and generation, deserves to be held in honored remembrance; and we are happy to have it reported to us this evening, that the work of preparing a fitting and just memorial of his life, and character, is entrusted to entirely competent hands, and that in due time, it will be ready for publication in our Historical Collections.

Resolved, That these Resolutions be entered at length on our records, in perpetual remembrance of the respect we bear for the memory of our deceased associate and friend, and of our grief at his death; and that an attested copy thereof be transmitted by the Secretary to the nearest relatives of Mr. Ward.

The acceptance of the resolutions was moved by Rev. G. D. Wildes and seconded by Prof. A. Crosby, and they were unanimously adopted.

MONDAY, OCTOBER 24. Evening meeting.

Vice President, A. C. Goodell Jr., in the chair.

Donations to the Library and Museum were announced.

Letters were read from—

Maine Historical Society, acknowledging the receipt of publications: Prof. S. F. Baird, of Washington, relating to the "Naturalist's Directory": H. L. Ordway, of Ipswich, on the habits of the Canker worm: Albert B. Russell and Theodore Atwill, of Lynn, accepting membership: Department of the Interior, Washington, giving notice of the transmission of books: A. L. Babcock, of Sherborn, relating to exchange of specimens: Dr. A. S. Packard Jr., of Brunswick, Me.; John W. Young, of Worcester; Miss Mary H. Coffin, of Newburyport; S. Lincoln, of Boston; S. J. Young, Librarian of Bowdoin College; Joseph Willard, of Boston; Wm. A. Smith, of Worcester, on business matters: James C. Ward, of Northampton, in reply to a communication containing the resolutions in memory of his father, the late G. A. Ward, Esq.

F. W. Putnam exhibited a skeleton of a Green Turtle, which had been prepared from a specimen lately presented by Francis Peabody, Esq., and explained the various parts of the skeleton, comparing it with that of a bird. He also spoke of the different sub-orders and families of Turtles as characterized by the skeleton, and exhibited a skeleton of the Chelys Matamata from the Amazon, which had been in the possession of the Institute for nearly thirty years, but had only recently been prepared for exhibition.

The Secretary presented, in the name of the Heirs of the late Perley Putnam, an autograph letter of General Lafayette, accepting the invitation to visit Salem in 1824, and made some remarks on the visit of Lafayette to this country in 1824—25.

The request of the "Picture Committee" of the National Sailor's Fair, for the loan of the portraits of John Rogers, Andrew LeMercier, Samuel Sewall, William Pinchon, Samuel Cooper, Benjamin Colman, Thomas Prince and Edward Holyoke, was referred to the Board of Directors.

WEDNESDAY, NOVEMBER 9. Stated meeting.

Vice President, A. C. Goodell Jr., in the chair.

F. W. Putnam proposed several amendments to the By-laws, which were adopted.

Solomon Lincoln Jr., of Salem, was elected a Resident Member. Edward S. Morse of Gorham, Me., and Edwin C. Bolles of Portland, Me., having been nominated by the Directors, were elected Corresponding members.

MONDAY, NOVEMBER 14. Evening meeting.

Vice President, A. C. Goodell Jr., in the chair.

Letters were read from the following:

Minnesota Historical Society, acknowledging the receipt of publications: Major Albert Ordway, 24th Mass. Infantry; Lt. John S. Allanson, 1st New York Engineers; Alex. Agassiz, of the Museum of Comp. Zoölogy; Alpheus Hyatt, of Cambridge, relating to the transmission of specimens: James C. Ward, of Northampton; E. M. Stone, of Providence, R. I., relating to the transmission of books: Prof. A. S. Packard, of Brunswick, Me.; J. S. Lewis, of Batavia, N. Y.; E. S. Morse, of Gorham, Me.; S. I. Smith, of Norway, Me., in relation to the publications.

F. W. Putnam read a communication from J. A. Allen of

Springfield, entitled "Notes on the habits and distribution of the Duck Hawk, or American Peregrine Falcon, in its breeding season, and description of its eggs," which was referred to the Committee on Publications.

Mr. Putnam presented, in the name of Rev. E. C. Bolles, of Portland, a collection of land and fresh water shells from Maine and New York.

Mr Bolles, who was present by invitation, being called upon, remarked that he felt like little more than a beginner in this department of conchology. He had been attracted to the study by the examination of the lingual ribbons of the land mollusks, organs remarkable for their beauty and regular structure, and exhibiting under the microscope fine specific characters. As yet there are but a few American students of these shells. In general, people are ignorant of the riches scattered about them in every forest and on every hill side. A snail is *only a snail* to almost everybody, and the common belief is that there is only one species and that unworthy of a serious man's attention. In Maine from which most of these specimens were brought, there are fifty species of land and fifty-four of fresh water mollusks. Most of these are forms peculiar to N. America. One, the Achatina lubrica is a cosmopolite, the same in both hemispheres, on islands and on continents. Some are analogues of foreign shells,—not facsimiles, but built on the same general plan. A few were evidently imported—carried by the accidents of commerce, as vermin and weeds have been, to make the grand tour of the globe. The islands of the Maine coast were early colonized. Sometimes old coins and carved stones are discovered there. There is another proof of European visits. The common snails of England still retain their rights of squatter sovereignty upon the soil. These shells have never been found far inland. They testify like the weeds which follow the pioneer to the great tide of nature's migration.

These specimens show us another great law of nature. Dissimilar as they are, all their differences lay in simple modifications of a simple type or plan. Beginning with Vitrina there is a loose transparent whorl of organized lime to protect the viscera of the mollusk. Through the flattened Helices to the turretted Achatina this whorl is twisted more or less closely, sculptured or plain, tinted or blanched, elevated or depressed, but in all cases reproducing the original plan in its structure. The animal exhibited the same fact. Animal and shell must be studied together. Here we begin to realize with what economy the Divine Wisdom worked. Out of a few simple substances and by touches of change almost microscopic in their minuteness the living vesture of the globe is made so various in its beauty and exhaustless in its forms.

The study of the anatomy of these mollusks is rendered somewhat difficult by the softness of their bodies. The most wonderful organ is the tongue or lingual membrane,—a rasp by which the creature secures its food. Each tooth of this rasp seems formed of the clearest glass. In some species there are over two thousand of these teeth upon the lingual organs. Under the microscope and especially by polarized light they form beautiful objects for examination. Mr. E. S. Morse, to whom the Natural History of Maine owes so much, has studied this matter scientifically and with fine results.

In short—Nature at our side everywhere offers us the choicest encouragement, whatever our particular tastes. The land repeats the wonders of the sea, and any association, like the Essex Institute, to study the lessons of both, is an association for mutual enjoyment, education and refinement in the knowledge of the great Creator.

The donations to the Library and Museum, received since the last meeting, were announced.

Charles Babbidge, of Salem, was elected a Resident Member.

MONDAY, NOVEMBER 24. Evening meeting.

The President in the chair.

Letters were read from the following:

Rev. E. C. Bolles, of Portland, Me.; Solomon Lincoln Jr., and Charles Babbidge, accepting Membership: A. S. Peabody, of Cape Town, Africa; C. H. Jones, of Sun Prairie, Wisc., relating to the transmission of specimens: A. R. Burton, of Littleton, N. H.; William Muir, of Fox Creek, Mo., relating to exchanges: Rev. James Hubbert, of Toronto, C. W.: Prof. A. E. Verrill, of Cambridge; Charles W. Felt; Robert Hamlin, of Bennington, Vt., on business matters.

F. W. Putnam read letters from George C. Huntington, of Kelly's Island, Ohio, giving an account of the "Red bug" of that Island, specimens of which were presented to the Institute by Mr. Huntington. Mr. H. stated that the insect was, as far as he could learn, found only on Kelly's Island. It is called the "Red bug" on account of its bright crimson color when living. It is so minute as to be hardly visible to the naked eye, and from its habit of penetrating beneath the skin, at the elbow joint, under the arms and other tender places, is very annoying to persons of delicate skin, especially to women and children; of late years however, it has been discovered that alcohol applied to the part affected will kill the insect and allay the eruption caused by it. Whence this insect comes, or where it goes, is still a mystery. They do not propagate while under the skin. In many of its habits it is similar to the "Jigger" of the Southern States, and it is thought by most persons to be the same insect, but by its size and structure this is at once disproved. Mr Putnam thought that the insect was allied to the Louse *(Pediculus)* and, as far as he could ascertain, it was as yet undescribed.

William P. Upham presented in behalf of Mrs. Martha Lee late of Manchester, an old Journal kept by Benjamin Craft during the siege of Louisburg in 1745, with letters written by him at that time; also a Journal kept by Eleazer Craft in the Revolutionary war, at the period of the surrender of Burgoyne, which was presented by Mrs. A. H. Trask of Manchester.

After some remarks upon the subject by A. C. Goodell Jr., and Rev. G. D. Wildes, the communication was referred to the Committee for publication in the Historical Collections.

The Secretary presented in the name of S. H. Phillips, a portrait of President William H. Harrison, painted by Abel Nichols Jr., of Danvers, who visited North Bend on the Ohio, for this purpose, during the Presidential campaign of 1840.

The chair made some remarks upon the events connected with this campaign, and mentioned several incidents illustrative of the character of the late President.

Two very handsome and large specimens of sponge collected from the piers of Beverly bridge, in the channel of the river, at about ten feet below low water mark, were presented by Rev. A. B. Rich of Beverly, who stated that these specimens exhibited, in his opinion, the two extremes of the species, as he had other specimens in his collection from the same locality, having intermediate forms.

Mr Putnam spoke of the structure of sponges and the various opinions of Naturalists as to their proper affinities, some holding them to be plants and others the lowest form of animal life; to the latter opinion he was strongly inclined.

R. S. Rantoul stated that the War Department had caused surveys to be made for one or two new forts, within the limits of our County. One of these is at Beverly and is intended as a part of the defence of Salem Harbor; for this fort the name of "Hale" would be appropriate, in honor of Col. Robert Hale, a distinguished citizen of Beverly, in the last century; and if the other should be located in Ipswich, it might be designated "Fort Dennison," in respect to the memory of Col. John Dennison, formerly one of the most noted personages in that section of the county. On Mr. Rantoul's motion a committee, consisting of Messrs. Davis, Rich and Tuck, all of Beverly, was appointed to confer with other parties in relation to the naming of the proposed forts, should they be erected.

Mr. Rantoul called attention to the large number of valuable manuscripts that were daily sent to the paper mills, and trusted that all present would endeavor to rescue as many old papers as possible and have them placed on file at the Institute.

A. C. Goodell Jr., followed Mr. Rantoul, and hoped that all the friends of antiquarian research would endeavor to save the old manuscripts, books, papers, &c., especially those of the Ante-Revolutionary period, from the collectors of such articles for the paper manufactories.

G. D. Phippen mentioned that during the past season Mr. C. W. Felt had removed his establishment for the manufacture of the Type-setting and Justifying machine to this city. Much interest having been expressed in this machine, which bids fair to change the present mode of composition in the printing office, Mr. Phippen moved that a committee be appointed to invite Mr. Felt or his associates to give an account of the machine at some future meeting of the Institute; Messrs. Huntington, Phippen, Goodell and Kimball were appointed on said committee.

James Talant of Concord, N. H., and James Hubbert of Toronto, C. W., having been nominated by the Directors, were duly elected Corresponding Members.

WEDNESDAY, DECEMBER 7. Special meeting.
The President in the chair.

The chair announced that the object of the meeting was to listen to an explanation of the Type-Setting, Justifying and Distributing Machine invented by C. W. Felt of this city, and now in the course of construction at the manufactory on Bridge street. After some general remarks appertaining to the subject, a general explanation of the machine, and of the purpose of its various parts, and their mode of operation was given by Mr. Wm. G. Choate, and a more detailed description of particular parts of the machine by Mr. John B. Richards, and remarks were made in regard to the invention by Mr. A. C. Goodell, Jr., and Mr. James Kimball.

This machine, as its name imports, sets and justifies type, and also distributes. The setting is done by the manipulation of a key board. There are thirty-seven keys for setting the type, one for each letter and character of some one alphabet, or size of type. While other keys touched with the keys of the several letters, turn the letters into any required alphabet, or size of type. Thus there is an italic key, and a capital key, which touched with the key of any letter, turn that letter into a capital or an italic, &c. The mechanism is so arranged as to keep pace with the most rapid compositor. Consequently if the manipulation of a key board is the quickest method of communicating motion intelligently to mechanism, as is believed, then this machine will enable a compositor to set types as fast as in the nature of things it can be done. Some idea may be formed of the rapidity with which the machine may be operated from the example of printing telegraphic machines which are operated by a similar key board. Rapid operators can compose on these at the rate of 7500 ems an hour, which is seven and a half times as fast as a rapid compositor can set types by the old method. By the use of a certain series of combined letters, cast in single type, which Mr. Felt has invented and which are used in the machine, there will be a further gain of about one-third, thus bringing the capacity of the machine nearly if not quite up to 10,000 ems an hour in the hands of a quick and skillful operator. Besides setting the type, this machine spaces and justifies the line, as well, or even better than can be done by hand and also leads the matter. The operation of justifying which printers have usually pronounced impossible for machinery to accomplish, and which no other type setting machine does or attempts to do, is performed by the machine automatically, all that the operator does, being to touch a key when his line is full, which transfers the line into the justifying apparatus and puts it in motion. Nor does the justification take the time of the operator. It is performed while he is setting the next line.

Attached to the machine is a register as it is called, which makes a complete record of all the operations of the machine by punching holes in a strip of paper. The use of the register is in rësetting and distributing the matter. These strips of paper being placed in the machine, and the machine set in motion, it will automatically set and justify the same matter in the same or a different type at any future time. This will obviate the necessity and save the expense of sterotyping books. The distribution is also automatically performed by means of the register, or it may be effected by the key board, or by nicks in the type.

Besides this machine Mr. Felt has invented several very simple and ingenious applications of the principles of the machine to setting type by hand which will be of great value, especially in small offices, where the large machines will not be required.

On motion of James Kimball it was *voted*—That the

thanks of the Institute be presented to Messrs. Choate and Richards for their interesting and instructive remarks and explanations of the machine.

MONDAY, DECEMBER 12. Evening meeting.
The President in the chair

Letters were announced from:

New Hampshire Historical Society; Maine Historical Society; Massachusetts Historical Society, acknowledging the receipt of publications: A. L. Babcock, of Sherborn and A. B. Burton, of Bethleham, N. H., relating to exchange of specimens: Dr. Wm. Wood, President of the Portland Society of Natural History; Lt. J. S. Allanson, 1st New York Engineers; Prof. S. F. Baird, of the Smithsonian Institution, on business: Rev. Joseph Banvard, of the Worcester Society of Natural History; Rev. E. C. Bolles, of Portland, Me.; Prof. A. E. Verrill, of Norway, Me.; J. A. Allen, of Cambridge; W. H. Dall, of Chicago, Ill., relating to the publications.

F. W. Putnam read a letter from William Hoxie, of Newburyport, in which Mr. Hoxie stated that he had found the following birds breeding in Byfield Parish during the past season—*Scolecophagus ferrugineus* Sw. (Rusty Blackbird), *Myiodioctes canadensis* Aud. (Canada Fly-catcher) and *Antrostomus vociferus* Bonap. (Whip-poor-will).

George D. Wildes read a memoir of the late Captain William Nichols, of Newburyport, a noted Privateersman during the war of 1812 and one of the most enterprising and daring navigators of that period.

On motion of Mr. Goodell the thanks of the Institute were tendered to Mr. Wildes for his interesting communication, and a copy was requested for publication in the Historical Collections.

Mr. Putnam mentioned that in a collection of Reptiles received from J. A. Allen, of Springfield during the past season, there was a specimen of the *Celuta amœna* B. & G. (Worm Snake). Mr. Allen had for several years past been confident that he had seen this species near Springfield, but had never been able to secure a specimen before. The only notice of this snake having been found in New England is by Dr. Storer who states, in his "Report on the Reptiles of Massachusetts," that a single specimen was collected by Professor Adams in Amherst, Massachusetts. Several authors having doubted the identification of Storer's specimen, the present one from Mr. Allen places the species beyond doubt in the Massachusetts fauna. Several specimens of *Heterodon platyrhinos* Latr. (Hog-nosed Snake, or Blowing Viper) were also in the collection received from Mr. Allen.

Mr. Putnam made some remarks upon the nest of a mouse found in a barberry bush, near Swampscott, and presented by Edward J. Porter.

A. C. Goodell Jr. mentioned that the course of Lectures on Insects, their habits and structure, by F. W. Putnam, would be delivered under the auspices of the Institute as soon as the necessary number of tickets were subscribed for.

Donations to the museum and library were announced.

W. P. Martin, W. R. Cloutman and E. S. Attwood, of Salem, were duly elected Resident Members.

MONDAY, DECEMBER 24. Evening meeting.

The President in the chair.

Letters were read from:

Messrs. Silliman & Dana, of New Haven, Conn.; T. A. Cheney, of Havana, N. Y., relating to an exchange of publications: Asst. Surgeon A. S. Packard, jr., 1st Maine Infantry; Alpheus Hyatt, jr., of Cambridge; James G. Arnold, Librarian, Worcester Nat. Hist. Soc., in relation to the publications: Edwin Harrison, of Irondale, Mo.; Albert G. Browne, Treasury Department, Beaufort, S. C., relating to the transmission of specimens; G. F. Matthew, of St John, N. B.; G. W. Tryon, jr., of Philadelphia, relating to the Naturalists' Directory: Prof. L. Agassiz, Director of Museum, Comp. Zoölogy; W. Barry, Sect'y Chicago Historical Society, acknowledging the receipt of publications: Rev. E. C. Bolles, of Portland, Me.; W. A. Nason, of Chicago, Ill.; W.

H. Dall, of Chicago, Ill.; B. O. Peirce, of Beverly; R. Kennicott, Sect'y, Chicago Acad. Nat. Science; Dr. J. Bernard Gilpin, of Halifax, N. S.; W. A. Smith, of Worcester; N. Paine, of Worcester, on general business.

F. W. Putnam read a communication from D. M. Balch, "On Native Grapes." In this paper Mr. Balch gives the results of his analyses of the following varieties of grapes grown in this vicinity, viz: the Delaware, Hartford Prolific, Concord, Adirondac, Allen's Hybrid, Union Village, Clinton, Alvey (Hagar), Franklin, Rogers' Hybrids Nos. 1, 3, 4, 9, 15, 19, 22, 30, 33, and 41.

From these analyses, native grapes would seem to be divided into three classes: 1st, those in which the proportion of acid and sugar are well balanced, as the Delaware, Rogers' Nos. 4 and 15, Allen's Hybrid, &c.; these should make good wine. 2d, those in which the acid is deficient, as in the Adirondac, Hartford, &c. 3d, those in which the great excess of acid overpowers all else, and renders the fruit nearly uneatable; such are the Clinton, Franklin, &c. The paper also contained several important practical remarks upon the culture of the grape in our climate. On motion of Mr. Putnam the communication was referred to the Publication Committee.

Mr. Putnam stated that, since the last meeting, he had ascertained that Mr. Samuels, in his report on the Mammals of Mass., mentioned that the White-footed, or Deer Mouse, *Hesperomys leucopus*, builds its nest in bushes, and he therefore presumes that the nest presented at the last meeting by E. J. Porter, was that of this species of mouse. In reply to a question from the chair, Mr. Putnam gave a brief account of the winter nests of the Musk Rats.

Charles Davis, in behalf of the committee appointed at a previous meeting, submitted a report containing the recommendation of the Selectmen of Beverly, that the Fort which the Government proposed to erect in Beverly, be called Fort Hale, in memory of Col. Robert Hale, formerly of Beverly, which was adopted.

Mr. Davis exhibited a fragment of the shell fired from the "Alabama" into the "Kearsarge," and which wounded three men on board the latter steamer; also the only piece of the "Alabama" remaining above water, and which was taken from the leg of one of the crew of the "Alabama" by Surgeon's Steward G. A. Tittle of the "Kearsarge," a citizen of Beverly.

Donations to the Library and Museum were announced.

James P. Kimball, of New York and Felipe Poey, of Havana, Cuba, having been nominated by the Directors were elected Corresponding Members.

Miss Susan T. Boynton, of Lynn and Henry W. Putnam, of Salem, were elected Resident Members.

Additions to the Museum and Library during October, November, and December, 1864.

TO THE NATURAL HISTORY DEPARTMENT.

BY DONATION.

Allanson, Lieut J. S., 1st. N. Y. Engineers. Lignite from the Dutch Gap Canal.

Barker, George. Skin of a Coot from Lake Cupsuptic, Me.

Barrett, ——— Miss, of South Danvers. Salamander, Hair Worms, and two Insects from South Danvers.

Bolles, Rev. Edwin C., of Portland, Me. 44 specimens, 25 species of Insects and Spiders from Portland, Me., and Mohawk, N. Y. 43 species of New England Land and Fresh water Shells.

Brown Jr., Benj. 12 specimens, 11 species native Insects.

Brown, Daniel. Fresh specimen of Blue Heron, *Ardea herodias* Linn.

Dall, W. H., of Chicago. 9 specimens of Reptiles from the vicinity of Lake Goodwin, Marquette Co., Mich.

Day, Albert. Specimen of Scorpion.

Emerton, James H. 87 specimens of native Insects.

Emmerton, Ephraim. A Lizard enclosed in copal.

Emmerton, W. H. Specimen of Sphinx taken in Salem.

Hale, Henry. Specimen of Blue Sulphuret of Iron, from which metalic paint is made.

Harrison, Edwin, of Irondale, Mo. Specimen of the Walking Fern, *Camptosarus rhizophyllus* from Irondale, Mo.

Haskell, Joshua P., of Marblehead. Cases of Worms resembling the shells of helix from Wenham Pond. 46 species of native Shells.

Heath, John. A Birds' nest from Marlboro, Mass.

Horton, N. A. A specimen of very thin veneer.

Ives, John M. Fresh specimen of Bittern, *Botaurus lentiginosus*.

Jillson, S., of Feltonville. Part of a skeleton of a Bald Eagle.

Jones, C. H., of Sun Prairie, Dane Co., Wisc. A collection of 130 Fishes from Madison 4th Lake, Wisc.

Kimball, Ed. D. Crown Crane, *Balearica pavonina*, from W. Africa.

Lake. Chas. H., Mass Vols. A " Green Rose" from Little Rock, Ark.

Lander, Miss E. R. Specimen of Lead from West Hampton, Mass.

Larrabee, E. L. 4 specimens of the Silver-side, *Atherina notata* captured under Beverly Bridge.

Lefavor, Joseph. Fresh specimen of a Woodchuck found in Salem.

Maloon, Wm. Hornets' nest.

Museum of Compartive Zoölogy, Cambridge. 40 Specimens, 30 species of Fishes from Singapore, collected by Capt. W. H. A. Putnam.

Nelson, Sylvanus, of Georgetown, Mass. Fossils from Rock river, N. Y.

Norris, Chas. H. Mud Turtle, *Chelydra serpentina*, from Salem.

Osgood, Capt. Charles. Eggs of Pyrula from Coast of Brazil.

Osgood, John C. Specimen of Crystalized Salt from Atlantic Salt Company's Works, Bay City, Michigan.

Peabody, A. S., of Cape Town, Africa. 2 specimens of *Callorhynchus antarctica* from South Africa.

Peabody, Francis. Green Turtle, *Chelonia mydas*.

Phippen, G. D. 12 specimens, 4 species Insects from Salem.

Porter, Ed. J. Nest of the White-footed Mouse, *Hesperomys leucopus*, found in a barberry bush in Swampscott.

Putnam, Chas. A. 6 specimens of Frost Fish *Morrhua pruinosa*, from North River. 30 specimens of *Unio complanatus* from Spring Pond.

Putnam. F. W. Eggs of the Sheldrake, Merganser, and Black Duck, from Oxford Co., Me.

Rich, Rev. A. B., of Beverly. Two large and handsome specimens of Sponge from the piers of Beverly Bridge.

Safford, Joshua. Ore from the Hecksher Coal Mine.

Samuels, E. A., of Boston. Egg of Fish Hawk, *Pandion carolinensis* Bonap., from Maine.

Sanborn, Francis G., of Boston. 227 specimens, 69 species of Spiders collected in Massachusetts.

Shute, James G., of Woburn. 73 specimens, 14 species of Insects from Newbern, N. C.

Todd, Mrs. John E. A. Native Silver from Copiapo, Chilli.

Tracy, C. M., of Lynn. 2 specimens of *Desmocerus palliatus*, male and female. Butcher Bird, *Collyrio borealis*, from Lynn.

Wheatland, Henry. Skull of a Musk Deer. Collection of Flower seeds from California.

Wheatland, Capt. Richard. Specimen of Tobacco grown in Salem.

BY EXCHANGE.

Babcock, Amory L., of Sherborn, Mass. 45 Skulls, mostly of native birds. Part of skeleton of a Cannibal fish from Surinam. Several Minerals from Kansas. Insects and Spiders from Sherborn.

Packard Jr., Dr. A. S., Brunswick, Me. Skulls of Seal, Esquimaux Dog, Black Bear and Squirrel, 13 specimens, 8 species of Fishes, 1 Frog, 16 species of Crustaceans, 200 specimens, 60 species Fossil Shells from the drift, 15 specimens, 3 species dried Echinoderms, from Labrador.

TO THE HISTORICAL DEPARTMENT.

BY DONATION.

Allanson, Lt. J. S., 1st N. Y. Engineers, Bermuda Hundreds. Fossil Wood from Dutch Gap Canal.

Andrews, Wm. P. A piece of one of the timbers of the oldest house in America, St. Augustine, Fla. Shells from Fort Wagner.

Black, Ensign Nath. W., of the gunboat "Mahaska." Rebel Torpedo from St. John's River, near Jacksonville, Fla.

Bolles, Rev. Edwin C., of Portland, Me. Rebel Envelopes. Several Postage Stamps, Foreign and American.

Brooks, Henry M. 7 City Checks for 5—25 cents from N. Y. & N. J.

Browne, Albert G., of Beaufort, S. C. Lime blocks from the oldest house in America, St. Augustin, Fla.

Cloutman, Wm. R. A Pike taken from the Chinese rebels by Gen. Ward. Japanese Custom House Receipt. Chinese Coin. East Indian Copper Coin.

Damon, ———, of Marshfield, Mass. Piece of Shell from the "Tennessee". Piece of Wood from the stern post of the "Brooklyn."

Davis, Chas., of Beverly. Photographs of the new Chapel of the Baptist Society in Beverly and of the first Pastor of the society.

Emerton, J. H. Cincinnati Token. Chinese Cash.

Emilio, Capt. Louis. Several Fuses of different kinds.

Fry, ———. Two Spindles from the railing on the top of a pew in the old East Church, built in 1718.

Miller, Fred. L., Ass't. Eng. U. S. S. Kearsarge. A piece of the shell fired by the "Alabama," and which wounded three men on board the "Kearsarge."

Ordway, Major Albert, 24th Mass., Infantry. 245 Foreign and American Coins.

PERKINS, GEO. Spoon made of bone by an American prisoner confined in Dartmoor prison during the war of 1812.

PHILLIPS, S. H. Portrait of W. H. Harrison, painted by Abel Nichols of Danvers.

PUTNAM, CAPT. W. H. A. Native Sword from Java.

ROPES, NATH., of Cincinnati, Ohio. 250 specimens, 108 kinds of Western Tokens.

ROPES, TIMOTHY. Image from ancient Thebes.

SMITH, WARREN A. Confederate 10 cts. Postage Stamp.

STONE, REV. EDWIN M., of Providence, R. I. A portion of the Cotton from the bale on which David Crowley of Providence floated and was saved from the burning "Lexington" Jan. 13, 1840.

TREADWELL. CAPT. W. A., 14th N. Y. Artillery. Confederate Pass taken from the body of Lt. J. B. Gayle, C. S. A., killed at Spottsylvania, C. H., May 14, 1864. Spur from the boot of a rebel killed at Bull Run.

WHEATLAND, H. Two antique Powder Horns.

TO THE LIBRARY.

BY DONATION.

BRIGGS, WILLIAM. Dedication of Forest Dale Cemetery, pamph. 8vo, Holyoke, 1862. 2d An. Rep. of New England Freedman's Aid Society, pamph. 8vo, Boston, 1864.

BROOKS, HENRY M. Roscobel, or the compleat history of the most miraculous preservation of Charles II., 1 vol., 12mo, London, 1725. An. Rep. of Adj. Gen. of Mass., pamph. 8vo, Boston, 1862. Ash's Grammatical Institutes, 1 vol., 16mo, Worcester, 1785. Boston Almanac 1861, 1 vol., 16mo. Brownlow's Knoxville Whig, several numbers. The Hive, vol. 1, 16mo, Salem, 1828—9. 22 Pamphlets.

CHASE, GEORGE C. Friend's Review, 20 Nos., Philadelphia, 1864.

DAVIS, CHARLES, of Beverly. The Alabama and the Kearsarge by F. M. Edge, 8vo, pamph., London, 1864.

DROWNE, CHARLES, of Troy, N. Y. Annual Register of the Rennselaer Polytechnic Institute, 1864–5, 1st term, 8vo, pamph., Troy, 1864.

FARNUM, JOSEPH. The Brunonian, edited by Students of Brown University, 1 vol., 8vo, Providence, 1831.

GILPIN, J. B., of Halifax, N. S. Bernard's Lecture on the Sable, Darby's wreck of the Arno, &c., 12mo, Halifax. 1858. Transactions of the Nova Scotian Institute of Natural Science, vol. I, pt. 1, and vol., II, pt. 1, 8vo, Halifax, 1863–4.

GREEN, SAMUEL A., of Boston. Radicalism in Religion, Philosophy and Social life, four papers from Boston Courier, 1 vol., 12mo, Boston, 1855. Catalogue of Lawrence Academy, Groton, 8vo, pamph., 1864.

Correspondence between Webster and Hulseman, 8vo, pamph., London, 1851. 71 Pamphlets.

IVES, HENRY P. Trial of the Murderers of Mr. White, 8vo, pamph., Salem, 1830.

KILBY, WILLIAM H., of Eastport, Me. Annual Report Adj. Gen. of Maine, for 1863, 8vo, 1 vol., Augusta, 1863.

KING, HENRY F. Hayden's Science and Revelation, 1 vol., 12mo, Boston, 1852. Barrett, the Golden Reed, 1 vol., 12mo, New York, 1855. Wilkinson on War, Cholera, and the Ministry of Health, 1 vol., 8vo, Boston, 1855. A Portrait of Swedenborg, &c., 8vo, pamph., Boston, 1854.

KING, MISS SARAH, of Danvers. Tull's Husbandry, 1 vol., 8vo, London, 1750. The Gentleman's Jockey, 1 vol., 8vo, London, 1683. Bulkeley and Cummin's voyage to the South Seas, 1 vol., 12mo, London, 1757. Alingham's Geometry, 1 vol., 12mo, London, 1714. United States Register for 1795, 1 vol., 16mo, Philadelphia, 1794. Salem Gazette for 1799, 1801, 1802, 3 vols., fol.

LANGWORTHY, I. P., of Boston. 43d, 48th, 49th, An. Rep. of American Tract Society, 8vo, pamph., Boston. American and Foreign Christian Union 32 Nos, and 64 Nos of the Christian World.

LORD, N. J. Files of the Boston Daily Post for June, July, August and September, 1864.

MASSACHUSETTS—SECT'Y OF STATE. Massachusetts Public Documents for 1863, 4 vols., 8vo. Report of Ship Canal, 1864, 1 vol., 8vo. 21st Registration Report of Mass., 1 vol., 8vo. Adj. Gen. Report, Mass., 1863, 1 vol., 8vo.

NICHOLS, JOHN H. Several papers printed at Charleston, S. C., during the year 1864.

ODELL, CHARLES. Collection of Almanacs.

SALEM, CITY. 277 Pamphlets,—principally Town Reports.

SIBLEY, JOHN L., of Cambridge. Catalogue of Harvard University, 1864–5, 12 mo, pamphlet.

SMITH, MRS., NATH'L., of Pembroke. Perin's Meditations, 1 vol., 16mo, Boston, 1709.

SPARKS, JARED, of Cambridge. Life of John Ledyard, by J. Sparks, 1 vol., 12mo, Boston, 1864.

STICKNEY, M. A. 40 Pamphlets. Introduction to Latin Grammar, 1 vol., 12mo. Exeter, 1794.

STORY, AUGUSTUS. Herald of Freedom, Railway Times and The Independent, files for several years. 24 Pamphlets.

TUCKER, JONATHAN. A collection of Manuscripts from the estate of the late G. Tucker.

UPTON, JAMES. Littell's Living Age, vols., 24, 25, and 26, 3d series, 8vo, Boston, 1864.

UNITED STATES—DEPARTMENT OF THE INTERIOR. Reports of the Pacific Railroad, vols. X and XI. 4to. Japan Expedition Report, vols. II and III. 4to.

VALENTINE, B. E. Catalogue of Haverford College, 1864–5, pamph., 12mo, Philadelphia.

WARD, CHARLES. Journal of Commerce Jr., files for July, August, September and October, 1864.

WARD, JAMES C., of Northampton. The Journal and Letters of Samuel Curwen, edited by George A. Ward, 4th ed., 1 vol., 8vo, Boston, 1864.

WILDES, JAMES H., of San Francisco, Cal. Maps of the Public Surveys of California and Nevada, 1863.

BY EXCHANGE.

ALBANY INSTITUTE. Transactions, vol. IV, 8vo, Albany, 1858–64. Transactions of the Society for the Promotion of the Useful Arts, vol. IV, pt. II, 8vo, Albany, 1819.

AMERICAN ACADEMY OF ARTS AND SCIENCES. Proceedings, vol. VI, pages 97 to 340 inclusive.

AMERICAN GEOGRAPHICAL AND STATISTICAL SOCIETY. Proceedings, vol. II, Nos. 3 and 4, 8vo, pamph., New York, 1864.

CANADIAN INSTITUTE. The Canadian Journal for Sept. and Nov. 1864.

DARTMOUTH COLLEGE LIBRARY. Catalogue of Dartmouth College, 1864–5, 8vo, pamph.

EDITORS. The American, Vol. I, No. 1, December, 1864, Salem.
Essex Banner, Haverhill, Mass.
Florida Union, Jacksonville, Fla.
Haverhill (Mass.) Gazette.
Historical Magazine, New York.
Lawrence (Mass.) American.
Lynn Weekly Reporter.
Salem Observer.
South Danvers Wizard.
The Palmetto Herald, Port Royal, S. C.
The Reader, London, England.
Tuolumne Courier, Columbia, Tuolumne Co., California.

GILMAN, D. C., Librarian of Yale College. Catalogue of Yale College, 1864–5, 8vo, pamph., New Haven, 1864.

HARVARD COLLEGE LIBRARY. 130 Pamphlets, principally relating to the various New England colleges.

HARVARD COLLEGE OBSERVATORY. Safford on the Right Ascension of the Pole Star, 8vo, pamph., Cambridge, 1864.

IOWA STATE HISTORICAL SOCIETY. Annals of Iowa for October, 1864, 8vo, pamph.

Massachusetts Historical Society. Proceedings, 1863–4, 1 vol., 8vo, Boston, 1864.

Massachusetts Institute of Technology. Scope and Plan of the School of Industrial Science, 8vo, pamph., Boston, 1864.

Missouri State Horticultural Society. Proceedings at Ann, Meeting, January, 1864, 8vo, pamph. Proceedings of Missouri Fruit Growers' Association, for 1859, 8vo, pamph.

Montreal Society of Natural History. The Canadian Naturalist and Geologist for October, 1864.

New Hampshire Historical Society. Collections, vol. vii, 8vo, Concord, 1863.

Philadelphia Academy of Natural Science. Proceedings, Sept. and October, 1864, 8vo, pamph.

Portland Society of Natural History. Journal Vol. i, No. 1, 8vo, pamph. Proceedings, Vol, i, pages 97 to 128 inclusive.

Providence Athenæum. 29th Ann. Report, Sept. 1864, 8vo, pamph.

Publishers. North American Review, Oct. 1864.

Rhode Island Historical Society. One hundred and fifty pamphlets.

Young, Stephen J., Librarian of Bowdoin College. 10th Ann. Rep, of the Schools in Maine, 8vo, pamph. Ann. Rep. of Adj. Gen. of Maine, for years ending Dec. 1861, and Dec. 1863, 2 vols., 8vo. The Bowdoin Bugle, No. xiii, Nov. 1864. 12 Pamphlets relating to Bowdoin College.

Zoölogische Gesellschaft, Frankfurt, a. M. Der Zoölogische Garten, Vol. v, Nos. 2, 3, 4, 5, 6.

BY PURCHASE.

Draper's History of Spencer, 1 vol., 8vo, Worcester, 1860. Lincoln's History of Worcester with Hersey's Continuation, 1 vol. 8vo, Worcester, 1862. Meade's Old Churches, Ministers and Families of Virginia, 2 vols. 8vo, Philadelphia, 1857. Sewall's Ancient Dominions of Me., 1 vol. 8vo, Bath, 1859. Moore's Lives of the Governors of New Plymouth and Massachusetts Bay, 1 vol., 8vo, Boston, 1851. Thayer's Family Memorial, parts 1 and 2, 1 vol., Hingham, 1835. Plumer's Life of William Plumer, 1 vol., 8vo, Boston, 1857. Adams, John, The Works of, with a Life of the author, Notes, &c., by C. F. Adams, 10 vols. 8vo, Boston, 1850–6. Vallandigham, C. L., The Trial by a Military Commission, 1 vol., 8vo, Cincinnati, 1863. Field's Centennial Address and Historical Sketches, 1 vol., 12mo, Middletown, 1853. Millet's History of the Baptists in Maine, 1 vol. 12mo, Portland, 1845, Westcott's Life of John Fitch, 1 vol., 12mo, Philadelphia, 1857. Brick Church Memorial, 1 vol. 8vo, New York, 1861. Woodworth's Reminiscences of Troy from its Settlement in 1790 to 1807, 8vo, 1 vol., Albany, 1860. Barne's Settlement and Early History of Albany, 1 vol. 8vo, Albany, 1864. Humphrey's Life of Putnam, 1 vol., 12mo, Hartford, 1850. Hopkins, The Patriot's Manual, 1 vol. 12mo, Utica, 1828.

MONDAY, JANUARY 9. Regular meeting.

Vice-President Goodell in the chair.

Letters were read from:

E. S. Atwood of Salem; James S. Tallant of Concord, N. H., accepting membership: J. W. Young of Worcester, relating to the publications: L. C. Draper, Sec'y Wisconsin Historical Society; E. T. Cresson, Sec'y, Entomological Society of Philadelphia; W. H. Dall of Chicago, Ill.; J. E. Arnold, Libr., Worcester Society of Natural History; W. C. H. Waddell of the American Geogr. and Statistical Society; B. Westermann, & Co.; James E. Oliver of Lynn; Waldo Higginson of Boston, on Business.

F. W. Putnam exhibited several colored drawings by Dr. J. Bernard Gilpin of Halifax, N. S.

One of these was a winter scene, representing a moose feeding on the tender twigs of a young tree which it had pushed over for the purpose, by straddling the tree with its fore legs, bearing on it with its chest. Another drawing was probably that of an undescribed species of Trout from Nova Scotia. The remaining were figures of the "Nurse" or "Sleeper Shark," *Somniosus brevipinna* Le Su., taken from a specimen captured in seventy fathoms of water on Sambro Banks, and brought to Halifax in the winter of 1862-3. The specimen was eleven feet three inches in length. In the manuscript accompanying the drawings, Dr. Gilpin describes the stomach, small and large intestines of this shark as being formed of one large simple gut from the mouth to the anus, with hardly perceptible differences in the various parts. He also mentions that there was a single cœcal appendage. This shark is said to inhabit deep water, never appearing on the surface, and its habits are so sluggish as to allow of its being often captured with a cod line. The fishermen speak of it as voracious, and, at some seasons, troublesome about their nets. Dr. Gilpin also remarks upon the inaccuracy of the published figures, of this species, by Le Sueur, DeKay and Yarrell.

Mr. Putnam spoke of the importance and great value of such figures and observations as those made by Dr. Gilpin, and called attention to the articles on the habits of the

Herring by Dr. Gilpin, published in the Transactions of the Nova Scotian Institute of Natural Science.

Rev. G. D. Wildes presented, in the name of Mrs. John Forrester, a Chinese visiting card of the late D. Fletcher Webster Esq., used while Secretary of the American Embassy at China. Also a Hindostanee Poem, written on Palm leaves and supposed to be five hundred years old.

Donations to the Library and Museum were announced.

Daniel H. Mansfield, Charles Odell, and Charles B. Fowler, of Salem, were elected Resident Members.

MONDAY, JANUARY 23. Regular meeting.

The President in the chair.

Letters were read from:

E. S. Morse of Gorham, Me., accepting Membership; Hiram A. Cutting of Lunenberg, Vt.; William Wood & Co. of New York; A. Mather of Philadelphia, in relation to the publications: J. A. Allen of Cambridge; R. Kennicott, Curator, Chicago Acad. of Sciences; E. T. Cresson, Sec'y, Entomological Society of Philadelphia; C. W. Felt; C. P. Preston of Danvers; Wm. Prescott of Concord, N. H., on business.

Rev. G. D. Wildes presented, in the name of Elijah Haskell, an old Spanish spear head and a gun lock, which were found in the Castle of San Juan d' Ulloa, Mexico; also the eggs of Pyrula from the Delaware Breakwater.

Donations to the Library and Museum were announced.

The President, after some appropriate remarks on the life,character and public services of the late Edward Everett, submitted the following resolutions:

Resolved. That we desire to express, and to place upon our records, in perpetual remembrance, our profound admiration and respect for the life and character of the HONORABLE EDWARD EVERETT, a Corresponding Member of the Essex Institute, and to join in the tributes, everywhere so justly and in such large measure paid to his memory, his worth, and his deeds, as the great American Scholar, Orator, Statesman and Patriot, and our most illustrious citizen.

A. letter from the Hon. C. W. Upham, was read by the President as follows:

Salem, Jan. 23, 1865.

Hon. Asahel Huntington,

President of the Essex Institute.

DEAR SIR:

It is eminently proper for every literary and scientific association to participate in the honors paid to the memory of EDWARD EVERETT. I regret not to be able to be present at the meeting this evening.

An uninterrupted friendship, covering a period of more than forty years, frequent and long continued correspondence, and much personal intimacy, have given me opportunity to judge of his character. I can say, with the strictest truth, that every word of encomium in the various forms in which the universal public sentiment has been expressed on the occasion of his death, finds full support in my impressions and recollections. In the combination of his natural endowments, the circumstances of his education and history, and the uses to which he put his great faculties and advantages, he has always appeared to me without a parallel.

The warmth and tenderness of his heart, his devotion to offices of benevolence, and his calm moral courage, are the traits which ever most arrested my attention. He often encountered vehement hostility, and the tide of popular misunderstanding and misrepresentation sometimes threatened to overwhelm him, but he kept on his way patiently and quietly, never yielded to its power, or veered from the course marked out by his convictions of duty.

He has been the great teacher of his countrymen of two generations, constantly pouring forth from his wonderful resources of knowledge and genius, the most useful information and the noblest sentiments. An elevating influence has pervaded all the productions of his pen, and inspired his eloquence. He has pushed forward the intelligence, and stimulated the progress of society steadily for more than half a century. If his unrecorded acts of courtesy, kindness, and usefulness in the daily routine, and ordinary course of life should be made known by those who have experienced them, they would equal in amount the extraordinary accumulation of his public labors. No one, however humble, ever addressed him for information without receiving a prompt and considerate reply, no one ever sought his aid without receiving evidence of his kind endeavor to serve him. He was faithful, punctual, and true to every opportunity of usefulness.

The collection of his Orations and Addresses, when completed, will be found to possess the elements of value and interest that will secure for them a permanent place in the highest department of the literature of the language. They embrace a wider circle of knowledge and a greater variety of subjects, in a style of elegance, accuracy, and polish, than any other work, and will stand the test of time.

His career justifies, and his classic grace and dignity of countenance and mien would peculiarly adorn the most costly monument that a grateful people can rear. All coming generations ought to be enabled to behold the features and form of the American, who has wrought out, by a life of industry, duty and virtue, the most finished model of culture and civilization.

If a portion of the contribution, which wealth and patriotism are about to make to this object, could be expended in giving to the public, in a beautiful form, and at a cost within the means of the great body of the people, a full collection of his productions, of all kinds,—from his first academic efforts to his last expiring strains, pleading the cause of country and Christian charity in Faneuil Hall,—it would indeed be the grandest monument, and render his usefulness perpetual.

Yours, very truly,

CHARLES W. UPHAM.

Resolved. That the letter of Mr. Upham be entered at length upon the records, and that we cordially concur in its sentiments and estimation of the life and character of Mr. Everett.

Resolved. That an attested copy of these proceedings

be transmitted by the Secretary, to the family of Mr. Everett.

Rev. G. D. Wildes seconded the resolutions with appropriate remarks, and they were unanimously adopted.

Francis C. Webster of Salem, was elected a Resident Member.

MONDAY, FEBRUARY 6. Regular meeting.

Vice President Goodell in the chair.

Letters were read from:

Massachusetts Historical Society; Natural History Society of New Brunswick; Nova Scotian Institute of Natural Science; Corporation of Brown University, acknowledging receipt of publications: Prof. A. E. Verrill of New Haven, Ct.; W. H. Dall, of Chicago, Ill.; Thomas R. Drowne; James L. Oliver of Lynn; C. W. Felt; J. Colburn of Boston; S. D. Bell of Manchester, N. H., on business: J. A. Allen of Cambridge; Trübner & Co. of London, relating to the publications: W. W. Stuart of the Buffalo Society of Natural Sciences, on the exchange of specimens: Rev. E. C. Bolles of Portland, Me., transmitting specimens: Prof. Theo. Gill, Librarian of the Smithsonian Institution, giving the particulars of the destruction by fire of a portion of the Smithsonian Institution: Franklin B. Hough of Albany, N. Y.; Prof. D. C. Eaton of Yale College, relating to the Naturalists' Directory: William Endicott of Shanghae, China, accepting Membership: Mrs. Mary H. Nichols, presenting a portrait of her late husband, Dr. Andrew Nichols: William Everett of Boston, in reply to a communication containing the resolutions passed at the last meeting of the Institute in memory of his father.

Dr. H. Wheatland gave a brief account of the life and services of Dr. A. Nichols, who was one of the pioneers in the study of Natural History, in this vicinity; following immediately in the steps of the celebrated Rev. Dr. Cutler of Hamilton. His example and precept have done much for the promotion of those objects which we now possess and enjoy. He was active in the organization of the Essex County Natural History Society, and for the first twelve years its President. It is well, occasionally, to look back upon the days of our infancy, and call to mind those who have laid the ground work of the operations of the present day.

Messrs. J. M. Ives and F. W. Putnam, alluded to the various discoveries made by Dr. Nichols in local Natural History.

F. W. Putnam announced the donation of one hundred and thirty-five copies of "The Victoria Regia, or the Great Water Lily of America, by John Fisk Allen," from the author. This work was published in 1854. It is a folio, and contains sixteen pages of text, and six plates representing the flower of natural size, in several stages of its growth, the structure of the leaf, and the young plant. The Institute, is, by this donation, in possession of all the remaining copies of the edition, and the only source whence the work can be obtained. On motion of Mr. Putnam, it was

Voted. That the copies of the "Victoria Regia," donated by Mr. Allen, be sold at a price not less than ten dollars per copy, or exchanged for works, equal in value, on Natural History and Horticulture, and that all monies received from this source be expended in the purchase of works on Natural History and Horticulture; and that all books received as above be placed in the Library of the Institute as donations from Mr. Allen.

Donations to the Library and Cabinets were announced.

Albert J. Lowd, Thomas R. Drowne and Benjamin Pearson, of Salem, were elected Resident Members. Winslow Lewis of Boston, was elected a Corresponding Member.

MONDAY, FEBRUARY 8. Quarterly meeting.

Vice President Goodell in the chair.

The following amendments to the By-Laws were adopted.

CHAPTER I. The following to be added: "Provided, however, that any member may, in lieu of the annual assessment, pay the sum of thirty dollars to be added to the funds of the Institute, the annual interest thereof to be considered as the payment of the annual assessment of said member."

CHAPTER IV. Lines 4th, 5th, 6th and 7th to be changed so as to read; "No specimen shall be taken from the rooms except by permission of the Committee of the Department to which it belongs, upon a written application made to the Secretary or Superintendent."

Amos Noyes of Newburyport, John Kinsman, Frederick Lamson of Salem, were elected Resident Members. James C. Ward of Northampton, was elected a Corresponding Member.

MONDAY, FEBRUARY 20. Regular meeting.

Vice President Goodell in the chair.

Letters were read from:

Prof. F. Poey of Havana; Messrs. Trübner & Co. of London; Prof. A. E. Verrill of Yale College; J. A. Allen of Cambridge, relating to the publications: R. Kennicott, Curator, Chicago Acad. Nat. Science; Lt. Col. Ordway, Bermuda Hundred, Va.; A. L. Babcock of Sherborn, notice of the transmission of specimens: Prof. Poey of Havana; Dr. Winslow Lewis of Boston; F. C. Webster, accepting membership: Lyceum of Natural History of New York; Mass. Historical Society; Boston Society of Natural History, acknowledging the receipt of publications: New York Chamber of Commerce, giving notice of the transmission of books: N. Bouton of Concord, N. H., in reply to questions respecting the Records of the Conventions, at Exeter, 1774—'75. Dr. S. A. Green of Boston, J. S. Appleton of Boston, on business: H. G. Jones, Corresp. Sec't. Penn. Historical Society, relating to the exchange of publications of the State of Pennsylvania.

Mr. Putnam read a communication from J. A. Allen, entitled, "Notice of a Foray of a colony of *Formica sanguinea* Latr, upon a colony of a *black species of Formica* for the purpose of making slaves of the latter." Referred to the Publication Committee.

F. W. Putnam made a few remarks upon the development of the fins of fishes, and the subsequent absorption of certain fins in some species.

He had lately examined young specimens of *Achirus lineatus* Cuv., and had discovered that they possessed *pectoral* fins, which were situated very near the opercular openings and composed of four well developed rays. In two specimens, which were nearly three inches in length, the pectorals were perfectly developed, except on the left side of one specimen where no fin could be traced. In another specimen, about four inches in length, both pectorals were present. A number of larger specimens were without pectoral fins which has been considered as the normal condition of the species of the genus Achirus.

Donations to the Library and Museum were announced.

Voted. That this meeting be adjourned to Tuesday evening next.

TUESDAY, FEBRUARY 28. Adjourned meeting.

Vice President Goodell in the chair.

Hon. C. W. Upham read an interesting memoir of our late esteemed member, GEORGE A. WARD.

On motion of Dr. Wheatland, it was

Voted. That the thanks of the Institute be tendered to Mr. Upham for the highly interesting and valuable memoir of the life, character and services of our late member, George A. Ward, and that a copy be placed at the disposal of the Publication Committee for publication in the "Historical Collections."

MONDAY, MARCH 6. Regular meeting.

Vice President Goodell in the chair.

Letters were read from:

The Department of the Interior, giving notice of the transmission of twenty-eight volumes of Public Documents: S. P. Fowler of Danversport; N. Brown of Boston, relating to the publications; Natural History Society of New Brunswick, acknowledging the receipt of publications: T. A. Cheney of Havana, N. Y., relating to an exchange of publications: Prof. James Hall of Albany, N. Y., offering to complete the Institute's set of the Reports on the New York State Cabinet: W. W. Burrage of Boston, relating to the printing of Reports of the Classes of Harvard University; H. W. Putnam, City Point, Va.; William L. Welch, notice of transmission of specimens and photographs: Messrs. Hartman & Laich of Cincinnati, Ohio, on business.

Captain N. E. Atwood of Provincetown, being present, was called upon by the chair and gave an interesting account of several species of native fishes as observed by him—

The Cod fish of the Eastern coast of the United States is not an inhabitant of the waters south of Cape Hatteras; that cape being the southern limit of the species. The northern limit he could not state, though it was certainly far north of the Straits of Belle Isle. In regard to the Cod, on our eastern coast, being of one, two, or three species, he could not, as yet, decide, but judging from their habits alone there might be three species, and it was his greatest desire to devote the rest of his life to the solving of this and similar problems in ichthyology, which can only be done by a person spending a length of time at each of the fishing grounds on the coast; carefully collecting facts, examining and comparing a large number of specimens from each place. At present, all he could say was, that there was a great and constant variation in the habits and size of the Cod from the various fishing grounds on the coast. The Cod taken by troll lines in the Gulf of the St. Lawrence are much larger than those from any other place, while those taken by the hand lines are quite small. The largest Cod he had ever seen weighed one hundred and one-half pounds, and this specimen was taken near Provincetown. He had heard of others that were supposed to have weighed from one hundred and fifty to one hundred and seventy-five pounds, which had been captured in the Gulf of St. Lawrence near the shore. On the coast of Labrador, he had never seen a Cod that would weigh twenty-five pounds; larger specimens, however, were taken on the small banks in the Straits of Belle Isle, five miles and upwards from the shore, and on George's Bank the fish caught with hand lines average larger than from other localities.

Mackerel come into Provincetown harbor in the spring as early as the 15th of May, all of full size and with spawn, and are then known as large No. 3's. By the 28th of May the spawn is fully developed, and deposited by the first of June. About the first of July young Mackerel, not more than two and one-half inches in length, are abundant in the Bay. These young Mackerel, in the latter part of October, are about six inches in length, and he has caught and packed and sold them as "No. 4 Mackerel." They leave the coast earlier in Autumn than the older ones. The large Mackerel, which appear first, as before stated, are followed by the arrival of small ones, on our coast north of Cape Cod about the 15th of June. These are known in the market as "Blinks," and are from last year's eggs. "Tinkers," are of two years growth; "Half-Size," are three years old, those older are called "Large ones." When Mackerel arrive on the coast, being lean, they are all designated as "No. 3's," but as they feed and improve in condition they are called "No. 2's," and when fat, are marked "No. 1," provided that they are thirteen inches long: but if less than thirteen and over eleven, then they are "No. 2's" if fat; all under eleven inches are marked as "small No. 3's," whether fat or poor. Adult Mackerel of four years, or more, are the only ones which spawn on our coast, and they will not take the hook until they have deposited their spawn, when they become lean and voracious. Formerly it was supposed that the large Mackerel, which first appear in Provincetown harbor, had passed the winter in the mud, and many persons would not eat them owing to their supposed muddy taste. These large Mackerel go further north than the smaller ones, returning southward long after the others have left the coast, and are even captured in November and December in the vicinity of Provincetown. Capt. A. was convinced that the *Scomber grex* was the young of the *S. vernalis*, Mitchill, and not a distinct species.

After giving an interesting account of the various modes of capturing the Mackerel at different times of the year, Capt. A. alluded to the Bluefish and the changes which had taken place in its habits. This fish, which many years ago, was very abundant, and held in high estimation by the Aborigines of our country, wholly disappeared from our coast in 1764, and not a specimen was seen on the coast, so far as Capt. Atwood knew, for fifty years. In 1847 they returned to the North of Cape Cod in great abundance, and havè since been taken in large quantities in weirs and nets, and by the hook, near the shore. Now they avoid the shore, and, during the last year or two have kept in the Bay, where it is difficult to capture them, as they seldom take the hook, though, until recently, they were most voracious and game fish.

The Menhaden, which were formerly so great a pest to the fisherman, and considered only fit for manure, appear in vast numbers on the coast of Massachusetts during the summer, a little later than the Mackerel, and remain until late in the season. They are now a valuable source of income, being caught for the oil, which is pressed from them, and sold for $40 a barrel. The refuse, after the oil is extracted, is used as a fertilizer and commands a high price. The sides of these fish are also salted, packed in barrels, and sold at a good price for Mackerel bait. The Menhaden does not spawn while on our coast, and it is only in the few, which have been driven into the rivers and which do not leave the coast until December or January, that spawn has been found. In the month of August and September a few of the young Menhaden are seen in our harbors; but further south, along the coast of Virginia, the young are seen in countless millions, and in heavy storms are driven on shore and left to die. On the coast of Virginia, small Menhaden appear after the large ones have left for the North. From the fact that the Menhaden which visit us during the summer, are either of a large and uniform size, or quite young, this species is supposed to attain its growth in a single year.

On motion of Rev. Mr. Wildes, the thanks of the Institute were voted to Capt. Atwood, for his interesting remarks.

Donations to the Library and Cabinets were announced.

John Dixey, Joseph B. F. Osgood, and Edward H. Knight, of Salem, were elected Resident Members.

MONDAY, MARCH 20. Regular meeting.

Vice President Goodell in the chair.

Letters were read from:

Lyceum of Natural History of New York, acknowledging the receipt of publications: Boston Society of Natural History, acknowledging the receipt of a collection of plants, collected in Zanzibar, by Caleb Cooke: Amos Noyes of Newburyport, accepting Membership.

Donations to the Library and Museum were announced.

A. C. Goodell Jr. read a paper entitled the "Cavalier and the Puritan." This will be published in a separate form.

John Daland of Salem, and Eben F. Stone of Newburyport were elected Resident Members.

Additions to the Museum and Library during January, February, and March, 1865.

TO THE NATURAL HISTORY DEPARTMENT.

BY DONATION.

ALLEN, FRANCIS R., Hamilton, Fresh specimen of Bald Eagle, *Haliætus leucocephalus*, shot in Hamilton, Jan. 20.

ALLEN, J. A., Cambridge. Specimens of a Red and of a Black species of Ant, and the pupa of the Black species; taken from an army of the Red species, Springfield, Mass., July 30, 1864.

BABBIDGE, CHAS. H. Chalcopyrite from Cheticamp, N. S.

BABCOCK, A. L., Sherborn, Mass. 9 specimens, 7 species, Insects; Skin of *Sciurus hudsonius;* skeletons and parts of skeletons of 3 species of native Birds; specimen of *Anodonta fluviatilis*, from Sherborn.

BUTTRICK, S. B. Portion of the jaw of a Porpoise. Water Beetle, *Dytiscus sp.*, from South Salem.

BOLLES, REV. E. C., Portland, Me. 7 specimens of *Anodonta edentula* Lea, *A. Ferussaciana* Lea, *Unio iris* Lea and *U. calceolus* Lea from Milwaukee, Wis. 3 specimens *Mya arenaria* from the Postpleiocene at Gardiner, Me. 3 specimens *Coleoptera* from Africa.

CARY, GEO. A. 2 specimens of *Tellena* from Turk's Island.

CHAMBERLAIN, JAMES A. 2 Holothurians, dry.

CHICAGO ACADEMY OF NATURAL SCIENCES, Chicago, Ill. Skins of 8 species of Mammals and 64 species of Birds from the West and North. 36 species, 51 specimens of Western Bird's eggs.

DAVIS, CHARLES, Beverly. Egg of an African Ostrich.

Harrington, Augustus, North Becket. Crystal and massive Emery, Iron ore and Margarite, from Chester, Hampden Co., Mass.

Haskell, Elisha. Eggs of *Pyrula* from Delaware Breakwater.

Hatch, Chas. F. Flying fish, *Exocœtus sp.*, from off the mouth of the Amazon. Centipede from near Parahiba River, Brazil.

Heath, John. Larva of *Papilio Turnus* Linn. from Lynnfield.

Heath, N. 5 Insects from Salem.

Higbee, Chas. H. Malachite from Africa.

Hoffman, Capt. Chas. Lizard from Bissao, W. C. Africa.

Kezar, Walter A. Wood perforated by Teredo, from Pensacola, Fla.

Kneeland, Cyrus A., Topsfield. Living specimen of the Saw-whet Owl, *Nyctale acadica*, captured in Topsfield.

Lovett, Edmonds. 3 species, 5 specimens of *Ophidians;* 3 species, 3 specimens of *Saurians;* 1 *Bird;* 6 species, 21 specimens of *Insects;* from the South West Coast of Africa.

Lowd, Mark. Fungus.

Mack, Dr. Wm. Larvæ of *Œstrus Bovis* from a cow.

Nelson, S. Augustus, Georgetown. Jasper from Winter Island.

Nichols, Stephen. Fungus.

Ordway, Col. Albert, 24th Mass. Inf't. Clay from "Dutch Gap Canal."

Palfray, Chas. W. Specimen of the Mocking Bird, *Mimus polyglottus* Boie, 13 years old.

Patch, W. H. H., Concord, N. H. Living Opossum, *Didelphys virginiana* Shaw, from Virginia.

Perkins, Ezra, Essex. Nest of Humming bird, *Trochilus colubris* Linn.

Pond, T. M., Framingham. Nest and eggs of the Meadow Lark, *Sternella magna* Swains, from Illinois.

Porter, Ed. J. Fossils and Minerals from Ohio. 10 Insects, 7 Crustaceans, from Essex Co. *Mantis sp.* from Washington.

Purdie, H. A., Boston. Several Spiders from Boston.

Putnam, F. W. Cochineal Insects, *Coccus cacti*, from Mexico. Specimen of Sapphire. Several species of Fishes from Mass. Bay. Claw of Lobster (malformation.)

Putnam, H. W. Infusorial Earth and Marl from near City Point, Va.

Quimby, Dr. E. H. Human embryo.

Roberts, J. W. Fresh specimen of the Great Gray Owl, *Syrnium cinereum* Aud., captured in North Salem.

Sanborn, F. G., Boston. Specimens of the Potter Wasp, *Eunemes fraterna*. Plum Weevil, *Rhynchænus Nenuphar;* Pine Weevil, *Curculio Pales*, from Mass.

Saunders, Miss Mary. Living specimen of the Acadian Owl, *Nyctale acadica*, captured in South Salem on Jan. 19.

White, Geo. M. 226 specimens, 75 species, Insects from Salem.

TO THE HISTORICAL DEPARTMENT.

BY DONATION.

ALLEN, J. F. Waterproof Japanese Coat made of paper.

BOLLES, REV. E. C., Portland, Me. One of the first stamped Envelopes issued in England, May 5, 1840. Designed by G. W. Mulready, R. A.

BUTTRICK, S. B. The Cockade worn by the late High Sheriff, Joseph E Sprague. Indian Relics from Ware's Beach, Marblehead, collected by JOHN W. BARTLETT.

CHAMBERLAIN, JAMES. 4 Foreign Postage Stamps.

COGSWELL, BRIG. GEN'L. A series of Photographic Views taken in Atlanta, Ga.

EDWARDS, CHAS. W., Serg't 2d Mass. Inf't. Piece of the Rebel Flag found flying at Atlanta on the capture of the city.

FAIRFIELD, CAPT. JAMES. 3 Coins from Uruguay ; 1 Coin from Buenos Ayres.

FELT, S. Q. Native Dress (Sarong) from Java. 5 Photographs of Sikhs and other castes of India.

FORRESTER, MRS. JOHN. Chinese Visiting Card of D. F. Fletcher.

GOODELL JR., A. C. Bill Head of John Hancock.

HASKELL, ELISHA. Gun Lock and an ancient Spanish Spear Head from the Castle of San Juan d'Ulloa, Mexico.

IVES, JOHN M. Wooden Images, carved by Alfred Bates, a soldier of the war of 1812.

KING, CAPT. H. F. Dutch Copper Coin (2 stivers.)

LOVETT, EDMONDS. Native Mat from the South West Coast of Africa.

PERKINS, EZRA, Essex. Indian Pipe.

PHIPPEN, NATH'L. $5 Bank Note of the United States Bank 1828.

PUTNAM, MRS. EBEN. Cane made from a timber used in the 2d and 3d house of the 1st Church.

RHODES, HENRY W. Rebel Uniform Button.

STEERS, JAMES L. Rebel Uniform Button.

SYMONDS, GEO. W. Piece of a Rebel Gunstock from the "Wilderness."

WILKINSON, MRS. ELIZABETH, Beverly. Irish Flax ; such as was used in the manufacture of linen cambric. Brought from Ireland in 1791.

TO THE LIBRARY.

BY DONATION.

ALLEN, CHARLES A., Cambridgeport. 1st and 2d Triennial Reports of Class of 1858 of Harvard, pamph.

ALLEN, JOHN FISKE. Boston Cultivator for 1862, 1863, 1864, 3 vols., 4to, Boston. 135 copies of the Victoria Regia, folio.

BALLARD, DAVID, Brunswick, Me. Bourne's Address at the Popham Celebration, Aug. 29, 1864, pamph.

Bolles, E. C., Portland, Me. Illustrations de Timbres Postes, J. B. Moens. 8vo, Liv. 1 to 9. Bruxelles, 1862.

Boston Public Library, Trustees of. 12th Annual Report, 1864, 8vo, pamphlet.

Brooks, Charles T., Newport, R. I. Carriers New Year's Addresses, Jan. 1, 1865.

Brooks, Henry M. Gospel of St. Mark, tr. and arranged by L. A. Sawyer, 1 vol., 12mo.

Brooks, M. C., James. Speech of J. Brooks in U. S. Congress, Dec., 1864, 8vo, pamph.

Burrage, Wm. W., Boston. Reports of Secretary of Class 1856 of Harvard for 1860, 1861 and 1865, 8vo, pamphlets.

Chase, George C. Friends' Review, 16 Nos.

Chase, George H. The Boatswain's Whistle, National Sailor's Fair, Boston, Nov. 1864, 1 vol., 4to.

Chase, Mrs. George H. The Sanitary Commission Bulletin, Nos. 1 to 32 incl., 8vo. The Sanitary Reports, Vol. 1 and Vol. 2, Nos. 1 to 18, 4to, Louisville, Ky, 1863. 30 Miscellaneous publications of the Sanitary Commission.

Choate, Wm. G. Rep. Adj. Gen. Mass. 1863, 8vo, 1 vol. Report of Commissioners of Agric. for 1862, 1 vol., 8vo. Manuals of Mass. Legis. for 1860 and 1861, 2 vols., 18mo. Midgley's Sights in Boston and Suburbs, 1 vol., 18mo. 45 Pamphlets.

Cole, Mrs. N. D. Salem Gazette for 1864, 1 vol., fol. Boston Daily Traveller for 1864, 2 vols., fol.

Colman, Benjamin. The act of Tonnage and Poundage and rates of Merchandise, 1 vol., 8vo, London, 1702.

Dawson, Henry B., Morrisania, N. Y. Correspondence between John Jay and H. B. Dawson, etc., concerning the Federalist, 8vo, pamph., New York, 1864.

Eastern Railroad, Directors. 30th Annual Report, 8vo, pamph.

Forrester, Mrs. John. The Overland Friend of India, files for 1858, 59, 60, 61, 62, 63, Serampore. Manuscript of a Hindostanee Poem on Palm leaves; supposed to be 500 years old.

Goodell Jr., Abner C. Lynn Directories, 1851, 1854, 1856, 1858, 4 vols., 16mo. Mass. Register, for 1858, 1 vol., 8vo. Boston Almanacs, 1861, and 1862, 2 vols., 16mo. 20 Pamphlets.

Green, Samuel A., Boston. Charleston Directory for 1862, 1 vol., 12mo. 8 Pamphlets.

Guild, R. A., Brown University. Jackson's account of R. I. Churches, 1 vol., 8vo. Ann. Cat. of Brown Univ. for 1856, 7, 8, 9, 60, 64. 16 Pamphlets.

Hall, James, Albany, N. Y. Account of Fossils of the Niagara Group, by J. Hall, 8vo, pamph.

Hanaford, Mrs. P. A., Reading. The Young Captain, a Memorial of Capt. Richard C. Derby, by Mrs. P. A. Hanaford, 1 vol., 16mo, Boston, 1865.

Holmes, John C. 25 Pamphlets.

Hutchinson, T. J. The Country Justice, by M. Dalton, 1 vol., fol., London, 1626.

Ives, Henry P. Robinson's Ancient History, 1 vol. Everett, L. S., Sacred Songs, 1 vol., 16mo. Jackson's Questions on the Lessons, &c., of the Church Service, No. 1, 1 vol., 16mo. Marshall's Public School Account Books. Worcester's Historical Atlas. 24 pamphlets. The American Publishers Circular for 1863.

Kilby, W. H., Eastport, Me. Address of Gov. Cony to Maine Legis. Jan'y 5, 1865, 8vo, pamph. 9th Ann. Rep. of Sect'y Maine Bd. of Agric., 1864, 8vo., pamph. Legis. Register of Maine, 8vo. pamph.

Langworthy, I. P., Boston. 152 Pamphlets, being Reports of various Charitable Societies, Minutes of Congregational Associations, &c.

Lewis, Winslow, Boston. Address at meeting N. E. Hist. Gen. Soc., 8vo., pamph., Boston, 1865. Report of Trustees of Mass. Gen. Hospital for 1864, 8vo, pamph.

Lord, N. J. Files of Boston Post for Oct., Nov., Dec., 1864.

Loring, George B. Files of Boston Post for 1863 and 1864, and Jan'y and Feb'y, 1865.

Miles, M., Lansing, Mich. Catalogue of Michigan State Agric. Coll. for 1864, 8vo, pamph. 2d Ann. Rep. of Sect'ry of Bd. of Agric. of Mich. for 1863, 8vo, pamph.

Nelson, Henry M., Georgetown. Ann. Rep. of Auditing and School Committees for 1865, 2 pamph., 8vo.

New York Chamber of Commerce. Annual Reports for 1858, 1859, 1860, 1861, 1862, 1863, 1864, 6 vols, 8vo. 30 Pamphlets, publications of the Chamber.

Nichols, George. Burnham's Historical Directory at Rindge, N.H., Nov. 14, 1861, 8vo, pamph. 1st An. Rep. of Discharged Sailor's Home, 8vo, pamph., Boston, 1863. Christian Enquirer for 1864, 1 vol., folio.

Packard Jr., A. S., Brunswick, Me. Synopsis of Bombycidæ of U. S. A., 8vo, pamph.

Paine, Nath'l, Worcester. Worcester Directory, 1865, 1 vol., 12mo.

Palfray, Charles W. Mass. Legis. Doc. for 1864, 2 vols., 8vo. 30 pamphlets.

Parish, A., Springfield. Report of School Comm. of Springfield, for 1864, 8vo, pamph.

Pease, Geo. W. Statement of the Account of Danvers from Feb'y 1864 to Feb'y 1865, 8vo, pamph.

Phillips, W. P. The Savannah Daily Herald and Savannah Republican.

RANTOUL, R. S. 24th, 25th, 26th Ann. Rep. of Mass State Bd. of Education, 3 vols, 8vo. Report of School Comm. of Boston, 1861. 1 vol., 8vo. Code Investigation 1860, U. S. Pub. Doc., 1 vol., 8vo. 7 Pamphlets.

SANBORN, FRANCIS G., Boston. Economical Entomology by F. G. S., 8vo, pamph.

SIBLEY, JOHN L., Cambridge. Ann. Rep. of Pres. and Treas. of Harvard College, 1863—4, 8vo, pamph.

STONE, BENJ. W. Valentine's Manual of the Common Council of New York. 1864, 1 vol., 12mo.

STONE, EDWIN M., Providence. 23d Ann. Rep. of the Ministry at Large in Providence, 8vo, pamph.

TAYLOR, SAMUEL L., Philadelphia, Ann. Rep. of Lib. of Penn. Hist. Society, 8vo, pamph., Philad., 1865.

UPHAM, WM. P. Flint's Agriculture of Mass., 1861, 1 vol., 8vo. Mass. Railroad Returns, 1863, 1 vol., 8vo. Ann. Rep. of Mass. Bd. of State Charities, 1 vol., 8vo, Boston, 1865.

WATERS, J. LINTON, Chicago. Ann. Rev. of Trade of Chicago in 1864, 8vo, pamph. Message of R. Yates, Gov. of Illinois, to the General Assembly, Jan'y 2, 1865, 8vo, pamph. 5th Biennial Report of Sup't of Public Instruction of Illinois, 1863-4, 8vo, pamph. Address of R. J. Oglesby, Gov. of Illinois, to Gen'l Assem., Jan'y 16, 1865, 8vo., pamph.

WHEATLAND, STEPHEN G. 22 Pamphlets.

BY EXCHANGE.

AMERICAN ANTIQUARIAN SOCIETY. Proceedings of Annual Meeting, Oct. 21, 1864, 8vo, pamph.

AMERICAN PHILOSOPHICAL SOCIETY. Proceedings, vol. IX, No. 72. 8vo, pamph. List of Members, 1865, 8vo, pamph.

CANADIAN INSTITUTE. The Canadian Journal for Jan'y, 1865.

EDITORS: American Journal of Science and Art, Jan. and Mch., 1865.
Florida Union.
Historical Magazine, for Jan. and Feb., 1865.
Haverhill Gazette.
Essex Banner, Haverhill.
Lawrence American.
Lynn Weekly Reporter.
The Reader, London, Eng.
The Newburyport Star.
The American, Salem.
Savannah Daily Herald.
Tuolumne Courier.
Salem Observer.

IOWA HISTORICAL SOCIETY. Annals of Iowa, No. 9, Jan., 1865, 8vo, pamphlet.

15

Massachusetts Historical Society. Collections, Vol. VII, 4th ser.

Massachusetts State Library. Report of the Librarian for the year ending September 30th, 1864, 8vo, pamph.

Minnesota Historical Society. Collections for the year 1864, 8vo, pamphlet.

New Brunswick Natural History Society. Jones on Ocean Drifts and Currents, 8vo, pamph. Bailey's Notes on Geol. and Botany of New Brunswick, 8vo, pamph. Bailey's Report on Mines and Minerals of New Brunswick, 8vo, pamph.

New England Historic-Geneological Society. Tercentenary celebration of the Birth of Shakespeare, April 23, 1864, 8vo, pamph. Lewis's Address before N. E. H. G. Soc., Jan. 4, 1865, 8vo, pamph. Tribute to the memory of Edward Everett, Jan. 17, and Feb. 1, 1865, 8vo, pamph.

New Jersey Historical Society. Collections, vol. vi, 8vo, Newark, 1864. Proceedings, vol. x, No. 1, 8vo, pamph.

New York State Library, Trustees of. Instructions on taking Census of New York, 1865, 8vo, pamph.

Philadelphia Academy of Natural Sciences. Proceedings for Nov. and Dec., 1864.

Philadelphia Mercantile Library Company. 17, 19, 21, 23, 24, 25, 26, 27, 29, 30, 31, 32, 33, 34, 35, 36, 38, 39, 41, 42 Annual Reports. Historical Sketch, 8vo, pamph. Special Report, April, 1863, 8vo, pamph.

Publishers. North American Review, Jan'y, 1865.

BY PURCHASE.

Saint John—The letters from an American Farmer, 1 vol., 12mo, Philadelphia, 1793. Sunderland—The testimony of God against Slavery, 1 vol., 16mo, Boston, 1838. United States, Foreign conspiracy against the liberties of, 1 vol., 16mo, New York, 1835. Rankin's Letters on American Slavery, 1 vol., 16mo, Boston, 1833. Bourne, The Book and Slavery irreconcileable, 1 vol., 12mo, Philadelphia, 1816. Paxton's Letters on Slavery, 1vol., 12mo, Lexington, Ky., 1833. Duncan's Treatise on Slavery, 1 vol., 12mo, New York, 1840. Clarkson's Essay on the Slavery and commerce of the Human Species, 1 vol., 12mo, Philadelphia, 1804. Buxton's Remedy for the Slave Trade, 1 vol., 12mo, New York, 1840. Thatcher's Indian Biography, 2 vols., 16mo, New York, 1832. Autobiography of Henry C. Wright, 1 vol., 12mo, Boston, 1849. Stuart's Memoir of Granville Sharp, 1 vol., 12mo, New York, 1836. Selections from the Writings and Speeches of William L. Garrison, 1 vol., 12mo, Boston, 1852. Memoir of Sebastian Cabot, 1 vol., 8vo, Philadelphia, 1831. Morse's system of Modern Geography, 1 vol., 8vo, Boston, 1814. Herrick.—A Genealogical Register of the name and family of Herrick, 1 vol., 8vo, Bangor, 1846. Secret Instructions of the Jesuits, printed verbatim from the London copy of 1725, 1 vol., 16mo, Princeton, N.

J., 1831. A. Bradford, New England Chronology, 1 vol., 12mo, Boston, 1843. Dowling, John.—The burning of the Bibles, defence of the Protestant version of the Scriptures, 1 vol., 16mo, Philadelphia, 1843. A Platform of Church Discipline of Synod at Cambridge in 1648, 1 vol., 12mo, Boston, 1808. A Book for New Hampshire Children in familiar letters, 1 vol., 12mo, Exeter, 1823. Edwards, Jonathan—A Treatise concerning the Religious affections, 1 vol., 12mo, Boston, 1768.

MONDAY, APRIL 3. Regular meeting.

Vice President Goodell in the chair.

Letters were read from:

Baron Osten Sacken, Russian Consul General, New York; John Akhurst, Brooklyn, N Y.; Philip S. Sprague, Quincy; Chas. M. Wheatley, New York; M. S. Bebb, Washington; Dr. H. C. Wood jr., Philadelphia; Dr. Wm. Wood, East Windsor Hill, Ct.; Wm. S. Vaux, Philadelphia; John Krider, Philadelphia; S. Jillson, Feltonville; D. G. Elliot, New York; T. McIlwraith, Hamilton, C. W.; John H. Thomson, New Bedford; Chas. J. Sprague, Boston; Wm. A. Smith, Worcester; B. Billings, Ottawa, C. W.; Dr. John L. LeConte, Philadelphia; Homer F. Bassett, Waterbury, Ct.; Nathl. Paine, Worcester; Sanborn Tenney, Cambridge; Isaac Lea, LL.D., Philadelphia; Geo. W. Peck, New York; Prof. James Hall, Albany, N. Y.; Dr. Theo. A. Tellkampf, New York; Dr. John W. Greene, New York; P. W. Sheafer, Pottsville, Pa.; H. A. Cutting, Lunenburg, Vt.; Prof. O. P. Hubbard, Hanover, N. H.; J. O. Treat, Lawrence; Dr. J. W. Robbins, Uxbridge; Thos. Barlow, Canastota, N. Y.; Dr. F. J. Bumstead, New York; W. W. Denslow, Inwood Station, N. Y.; Alex. Agassiz, Cambridge; H. F. King, Salem; D. M. Balch, Salem; Rev. David Weston; Worcester; G. F. & R. Matthew, St. John, N. B.; B. P. Mann, Concord; J. L. Sergeant, Philadelphia; Thomas Bland, Brooklyn, N. Y.; Chas. L. Blood, Taunton; Aliston Bacon, Natick; Prof. J. Leidy, Philadelphia; E. T. Cresson, Philadelphia; Jacob Ennis, Philadelphia; F. H. Hough, Albany, N. Y.; G. J. Bowles, Quebec, Canada; Prof. L. Agassiz, Cambridge; James Lemoine, Quebec, Canada; Geo. E. Brackett, Belfast, Me.; R. H. Brownne, New York; Ewd L. Graef, Brooklyn, N. Y.; Vincent Barnard, Kennett Square, Penn.; T. A. Greene, New Bedford; Prof. E. D. Cope, Philadelphia; J. T. Rothrock, McVeytown, Pa.; J. A. Allen, Cambridge; Alpheus Hyatt, Baltimore, Md.; Ed. S. Morse, Gorham, Me.; Prof. A. E. Verrill, New Haven, Ct.; T. A. Cheney, LL.D., Havana, N. Y.; John G. Hodgins, Toronto, C. W.; Rev. Chester Dewey, Rochester, N. Y.; John Macoun, Belville, C. W; R. P. Whitfield, Albany, N.

Y.; J. V. C. Nellis, Auburn; J. P. Haskell, Marblehead; John Orne jr., Salem; Rev. Joseph Banvard, Worcester; Prof. T. C. Porter, Lancaster, Pa.; Stephen Salisbury jr., Worcester; Prof. S. F. Baird, Smithsonian Institution; L. Satterlee, New York; Prof. J. L. Russell, Salem; William Couper, Quebec, Canada, relating to the publications: N. Brown, Boston; Wm. Stimpson, Corr. Sect., Chicago Acad. Nat. Sci.; A. L. Babcock, Sherborn; Rev. E. C. Bolles, Portland, Me.; Capt. N. E. Atwood, Provincetown, on business: Trustees of Boston Public Library, acknowledging the receipt of publications: E. F. Stone, Newburyport, accepting membership.

The Superintendent called the attention of the meeting to the fifty mounted specimens of birds and mammals on the table, stating that all the species were new to the collection and that they were the skins presented by the Chicago Academy of Natural Sciences and the Lyceum of Natural History of William's College.

Gilbert L. Streeter read a communication entitled, "Salem one hundred years ago," suggested by a perusal of a Dudleian lecture sermon by Rev. Thomas Barnard of Salem, delivered in May 1768, and printed in Salem by Samuel Hall, who about that time opened a printing office in Salem.

Remarks were offered by Messrs. Wildes, T. Ropes, and the chair, and on motion of the Secretary, a copy of the paper read by Mr. Streeter, was requested for publication in the Historical Collections.

Edward S. Thayer, Nathaniel Kinsman, Jonathan Ropes and Charles H. Pepper of Salem, were elected Resident Members.

MONDAY, APRIL 17. Regular meeting.

Rev. G. D. Wildes in the chair.

Letters were read from:

Prof. J. Henry, Sect. Smithsonian Institution; Isaac Lea, LL.D., Philadelphia, Pa.; Prof. C. A. Joy, Columbia College, New York; Dr. Wm Stimpson, Cor. Sect. Chicago Acad. Natural Sciences; Capt. A

Hyatt, Baltimore, Md.; Dr. H. C. Wood jr., Philadelphia; Rev. C. J. S. Bethune, Sect. Entomological Soc. of Canada; Dr. J. H. Salisbury, Cleveland, Ohio; W. W. Cary, Coleraine, Mass.; J. G. Sanborn, Cherryfield, Me.; Miss Julia H. Spear, Burlington Vt.; B. P. Mann, Concord, Mass.; Prof. Traill Greene, Easton, Pa.; D. G. Elliot, New York; S. B. Mead, Augusta, Ill.; Dr. J. W. Robbins, Uxbridge, Mass.; Dr. Asa Horr, Dubuque, Iowa; A. N. Prentiss, Lansing, Mich.; Dr. C. C. Abbott, Trenton, N. J.; J. R. Willis, Halifax, N. S.; John Kirkpatrick, Sect. Cleveland Acad. Natural Sciences; C. A. Emery, Springfield, Mass.; Dr. Ezra Michener, Avondale, Pa.; G. C. Brown, Mt. Holly, N. J.; J. F. Knight, Sect. Entomological Society of Philadelphia; A. L. Babcock, Sherborn, Mass.; Rev. J. E. Long, Hublersburg, Pa.; Isaac A. Pool, Chicago, Ill.; Frank Stratton, Natick, Mass.; C. F. Austin, Closter, N. Y.; Wm. W. Stewart, Custodian, Buffalo Society Natural Sciences; Prof. L. W. Bailey, University of New Brunswick; J. P. Lesley, Philadelphia; E. Tatnall jr., Wilmington, Del.; J. A. Allen, Cambridge; W. H. Niles, Cambridge; C. F. Hartt, Cambridge; O. H. St. John, Cambridge; Prof. T. C. Porter, Lancaster, Pa.; H. S. Babbitt, Asst. Sect. Ohio State Bd. of Agriculture; Prof. How, King's College, N. S.; D. Wilkins, Littleton. N. H.; Saml. P. Fowler, Danvers, Mass.; P. S. Sprague, Quincy, Mass.; J. D. Sergeant, Philadelphia Acad. Nat. Sciences; Wm. Gossip, Sect. Nova Scotian Institute of Nat. Science; Dr. H. M. Paine, Albany, N. Y.; H. A. Cutting, Lunenburgh, Vt.; B. S. Lyman, Philadelphia; Elihu Hall, Athens Ill.; Dr. Daniel Clark, Flint, Mich.; Dr. J. A. Meigs, Philadelphia, Pa.; Prof. D. S. Sheldon, Griswold College, Davenport, Iowa; R. Kennicott, Director, Museum Acad. Nat. Sciences, Chicago, Ill.; B. F. Mudge, State Geologist of Kanzas, Quindaro, Kanzas; Wm. S. Sullivant, Columbus, Ohio; Prof. F. Poey, Habana, Cuba; J. G. Arnold, Worcester; Dr. J. C. Draper, New York; Prof. S. S. Haldeman, Columbia, Pa.; Prof. A. E. Verrill, Yale College; C. M. Tracy, Lynn; Wm. S. Vaux, Philadelphia; Dr. Theo. Tellkampf, New York; E. T. Cresson, Corresp. Sect. Entomological Society of Philadelphia; Prof. Leo Lesquereux, Columbus, Ohio, relating to the Publications: C. M. Wheatley, relating to an exchange of specimens: Miss Anna L. Coffin, Newbury; C. M. Tracy, Lynn; Dr. Wm. Wood, Portland, Me.; N. Brown, Boston; Capt. N. E. Atwood, Provincetown, on business: G. A. Boardman, Milltown, Me.; W. W. Denslow, New York, offering to send specimens to the Museum: Albert G. Browne, Charleston, S. C.; J. C. Convers; Mrs. G. H. Chase; Blagden & Co., Boston, announcing the transmission of books for the Library: Smithsonian Institution, acknowledging the receipt of Publications: Charles Henry Pepper, accepting Membership.

Donations to the Library and Museum were announced.

Professor A. E. Verrill made some remarks upon the Iron ores found in the New England States and contiguous part of New York. Two kinds of ore were mentioned, the Hematitic and the Magnetic oxide. The bog ore found in our low lands resembled the Hematitic in its composition but is usually much inferior in quality.

Francis C. Butman, and Francis A. P. Rust, of Salem, were elected Resident Members. Professor Edward D. Cope of Philadelphia, Professor James Hall of Albany, and Baron Osten Sacken, Russian Consul General at New York, were elected Corresponding Members.

The Chair stated that at the last meeting, the news of the evacuation of Richmond had been received, and that on the morning of the Monday following the announcement of the surrender of Lee's Army was published to the Country. This day startling news of a different character was received: the death of the President of the United States, at Washington, on Saturday, April 15, at 7.20 o'clock in the morning, occasioned by a bullet wound from a pistol in the hands of an assassin on the evening previous. These events deserve a place on our records.

Robert S. Rantoul introduced a series of appropriate resolutions, which were unanimously adopted and ordered to be placed upon the records of the Institute. Dr. George B. Loring, on moving their adoption, paid an eloquent and deserved tribute to the memory of the late President. Professor A. Crosby followed Dr. Loring with suggestive remarks.

MONDAY, MAY 1. Regular meeting.

Vice President Goodell in the chair.

Letters were read from:

Chas. Stodder, Boston; Samuel R. Carter, Paris Hill, Me.; Wm. A. Haines, New York; Prof. T. S. Parvin, Iowa City; Asst. Surg. B. G. Wilder, 55th Mass. Vols.; Chas. Wright, Wethersfield, Ct.; Prof. Edw. Hitchcock, Amherst College; Dr. John Gundlach, Habana, Cuba; Tryon Reakirt, Philadelphia; Thomas Meehan, Editor of the Gardener's Monthly; S. I. Smith, Norway, Me.; John Bolton, Portsmouth, Ohio; S. D. Poole, Lynn; J. D. Parker, Steuben, Me.; Prof. Dana, Yale College; Isaac C. Martindale, Byberry, Pa.; Prof. D. S. Sheldon, Griswold College; Wm. S. Sullivant, Columbus, Ohio; Prof. H. A. Thompson, Otterbein University; G. F. Matthew, St. John, N. B.; Prof. A. E. Verrill, Yale College; S. B. Mead, Augusta, Ill.; Dr. J. Aitken Meigs, Philadelphia; W. J. Howard, Central City, Colorado; Dr. S. A. De Morales, Habana, Cuba, relating to the Publications: Wm. Wood & Co., New York; Wm. W. Stewart, Custodian, Buffalo Soc. Nat. Sciences, on business: A. M. Edwards, New York, announcing the formation of the American Microscopical Society in New York.

Donations to the Library and Museum were announced.

A large number of native plants, collected by Nathaniel Hooper and James H. Emerton, were placed on the table and were explained by Geo. D. Phippen, who had a few interesting remarks to make on each of the various species. Mr. Phippen thought that the opening of the flowers this year, was about ten days in advance of many previous years.

Messrs. Hooper and Emerton gave an account of the special locality of several of the rarer species. Mr. Emerton read a few notes relating to the time of flowering of a number of species of plants, the present season, and also as to the first appearance of several species of insects this spring.

F. W. Putnam stated that the Toads commenced spawning on the 16th of April. He then made some remarks, suggested by those of Mr. Phippen, upon the various theories regarding the origin of species.

Messrs. F. W. Putnam, Charles Davis, W. P. Upham and the Secretary were appointed a committee to nominate officers for the ensuing year, and report the same at the annual meeting.

Edward Dean and T. Francis Hunt, of Salem, were elected Resident Members. E. T. Cresson of Philadelphia, was elected a Corresponding Member.

WEDNESDAY, MAY 10. Annual meeting.

Vice President Goodell in the chair.

The reports of the Secretary, Treasurer, Superintendent, Curators and Committees were read and accepted. From these reports the following particulars may be specified.

The Society is in a good and healthy condition. The receipts from the assessments of the Resident Members have been greater than in any preceding year, which was also the case in regard to the sales of publications. During the year one hundred and fifty seven Resident and twelve Corresponding Members have been elected. Eight Resident Members have removed from the county, and the following have died during the year: Wm. B. Brown, Henry Hubon, Edward L. Perkins, Charles W. Swasey, Lucy Tréadwell, George A. Ward, Mary E. Wheatland, Samuel Webb, all of Salem. The sad intelligence of the decease of the following Corresponding Members has been received: Hon. Edward Everett of Boston, Mass., Professor Benjamin Silliman of New Haven, Conn., Carlton A. Shurtleff of Roxbury, Mass., and William B. Fowle of Medfield, Mass. Biographical notices of the deceased Members will be printed in the Historical Collections. The present number of Resident Members is five hundred and two, of Corresponding, one hundred and thirty-six.

Five field meetings have been held during the past season; at East Saugus, Wenham Pond in Beverly, Gloucester, Rockville chapel in South Danvers and Newburyport. These meetings have been largely attended, and a greater interest than at any previous season has been manifested.

Evening meetings have been held on the second and fourth Mondays of each month for the first part of the winter, and the first and third Mondays afterwards, at the rooms of the Institute, commencing in October and closing with the annual meeting in May. The large number attending these meetings calls for a more commodious meeting room at as early a day as practicable.

The Lecture Committee, having adopted the plan of having courses of lectures on special subjects and of an educational character, delivered to appreciative audiences, in lieu of the more extended courses of a miscellaneous character of former years, made arrangements with Messrs. Putnam and Tracy, who have taken the initiative, and the committee trust that this plan will be adopted in other branches.

F. W. Putnam, on the five Thursday evenings in March, delivered a course of lectures on "Insects, their habits and structure," at Lyceum hall, under the auspices of the Institute, which were very instructive and were well attended by highly appreciative and intelligent audiences. At the close of the course the following resolution, moved by Prof. A. Crosby and seconded by Gen. H. K. Oliver, was unanimously adopted:

"*Resolved:* That we express to Mr. Putnam our high appreciation of the valuable and interesting Course of Lectures he has just completed; and the personal thanks and obligations of our community to him for these labors in the cause of science and public improvement, especially in view of his generous appropriation of the greater part of the proceeds to the benefit of the Museum of the Essex Institute."

Cyrus M. Tracy of Lynn has delivered two of a series of eight lectures on Botany at the rooms of the Institute on the two preceding Saturday afternoons.

The Treasurer presented the following statement of the financial condition, for the year ending May, 1865.

GENERAL ACCOUNT.

Debits.

Athenæum Rent, half fuel, &c.	$535 53
Publications, $1001 25; collecting assessments, $23 10,	1024 35
Postage and Express, $76 90; Gas, $11 90,	88 80
Printing, $26 75; Stationery and Books, $33 03,	59 78
Sundries,	41 06
Historical account,	143 50
Natural History and Horticultural account,	59 34
Balance in Treasury,	12 27
	$1964 63

Credits.

Balance of last year's account,	7 04
Dividends Webster Bank, $60 00; Books sold, $180 19,	240 19
Sale of Publications,	809 40
Assessments,	908 00
	$1964 63

NATURAL HISTORY AND HORTICULTURE.

Debits.

Preservatives &c., $65 00; Specimens, $34 00,	99 00
Cases, $56 75; Bottles, $6 60,	63 35
Horticultural Exhibition,	53 10
	$215 45

Credits.

Horticultural Exhibition,	104 11
Dividends Lowell Bleachery,	40 00
" Portland, Saco & Portsmouth Railroad.	12 00
General Account,	59 34
	$215 45

HISTORICAL ACCOUNT.

Debits.

Binding, $100 00; Books, $98 50,	$198 50

Credits.

Dividends Naumkeag Bank,	17 00
Coupons Michigan Central Railroad,	38 00
General Account,	143 50
	$198 50

The Library is daily in receipt of additions. A large increase is consequent upon the exchanges that have been arranged with different societies, and editors or proprietors of historical and scientific journals, newspapers, &c.

The additions during the year, principally by donation or exchange, are:

Octavos and lesser fold,	500
Quartos 4, Folios 12,	16
Newspapers, Folios, (files)	88
Pamphlets and Serials,	1,500
	2,104

The above have been contributed by one hundred and twenty individuals and fifty-seven Societies, Editors of Journals and the various departments of the State and General Government.

The publication of the Proceedings and Historical Collections has been continued during the year: of the former, vol. III and Nos. 1, 2, 3, and 4 of vol. IV have been printed; of the latter, volume VI.

The annual Exhibition of Fruits, Flowers and Vegetables took place on Wednesday, Thursday and Friday, Sept. 21, 22, 23, and exceeded our expectations after the severe unprecedented drought. There were many plates of fine pears. The leading feature was the display of outdoor grapes, which was judged the finest ever exhibited in the state. The vegetables were particularly fine. The flowers, as usual, were very attractive and contributed essentially to the general appearance of the rooms.

The Historical Department has been increased during the year by the addition of two hundred specimens to the Ethnological section, a large number of valuable manuscripts and several engravings and portraits. The room given to this department is much crowded, rendering a proper display of the collection impossible and obliging the storage of many of the engravings and manuscripts for the want of accommodation.

Two hundred and thirty-seven donations, embracing eight thousand five hundred and three specimens, have been received for the Natural History Department during the year. The work in the various sections of the department has been carried on with good results, and several are in a forward state of arrangement. Catalogues have been commenced and in some of the classes the specimens are as far arranged as the limited supply of case room, jars and alcohol will allow for the present. We are under great obligations to Professor Verrill, of Yale College, for the identification and arrangement of the Polyps, and Acalephs. These classes have been largely increased by the valuable addition of several hundred specimens of East Indian corals, collected and presented by Capt. W. H. A. Putnam. By the kindness of Professor Verrill, and Mr. Alexander Agassiz, who had previously identified the collection of Echinoderms, we have the specimens belonging to the branch of Radiata so far identified that it is proposed to publish a catalogue of the collection at an early day. We are also indebted to Rev. E. C. Bolles, of Portland, for the identification of many of our native land and fresh water shells.

It is to be hoped that the work on the collection will not long be impeded by the present insufficient supply of cases and materials for the proper exhibition of the specimens. Much larger accommodations are required for the various departments of the Museum and these cannot be supplied to the extent desired without an addition to the present building by which, at least, three times the present amount of case room can be obtained. Not more than three-fourths of our specimens are now visible to the public, or of use for study, as many are stowed in kegs and cans in the cellar and in drawers and boxes in the hall. A partial supply of case room could be obtained by the construction of a few railing and table cases in the hall for the Insects, Fossils and Birds' nests and eggs. The cases for the pinned Insects are needed at once, for this valuable collection is being destroyed by its insect enemies, and until more room is given to it this destruction cannot be wholly prevented, even by the constant vigilance of the Curator

As the arrangement of the various classes is perfected large numbers of duplicate specimens are separated, which will be presented to such institutions and individuals as will use them for the advancement of science, in accordance with the rule adopted by the Institute regarding the distribution of its duplicates. Though a number of collections are now being packed for transmission to various persons and societies the following, only, have been sent during the past year.

To the Cabinet of Yale College:

40	species,	102	specimens,	of	Corals.
25	"	41	"	"	Echinoderms.
7	"	20	"	"	Sponges.
4	"	5	"	"	Tunicates.

To A. L. Babcock, of Sherborn, Mass.:
168 specimens of South American and
2 " " African Insects.

To the Chicago Academy of Science:
25 species, 110 specimens, of foreign Helices.

To the Museum of Comparative Zoölogy:
1 specimen of *Goniaster cuspidatus* Gray, from the West Coast of Africa.

To J. G. Shute of Woburn, Mass.:
28 species, 59 specimens of foreign Shells.

To Rev. E. C. Bolles, of Portland, Me.:
73 species, of several specimens each, of foreign and American land and fresh water Shells.

The following estimate of the number of specimens (exclusive of a large number of duplicates) in the various departments of the Museum presents a general view of the character of the collection at the present time.

HISTORICAL DEPARTMENT.

The section of Ethnology contains about 1400 specimens, illustrating the habits, costumes, war and domestic implements of the various races and nations.

In the section of Manuscripts there are a very large number of Manuscripts relating to our early civil and ecclesiastical history.

In the section of Fine Arts there are several hundred Portraits, Paintings and Engravings, many of which are of great historical value.

DEPARTMENT OF NATURAL HISTORY.

	Specimens.
Geological specimens, about	200
Minerals, 1896 specimens, of which 196 are from Essex County, ...	1896

FOSSILS.

Radiates,	186 species,	250 specimens.			
Mollusks,	1108 "	2000 "			
Articulates,	20 "	50 "			
Vertebrates,	90 "	120 "			
Plants,	135 "	200 "	*Fossils,*	.	2620

RECENT.

Plants, about 5000 species, native and foreign, among which are nearly all the species found in Essex County, a number of specimens of wood, and a large number of seeds· &c., in all about 6500. *Plants,* . 6500

Sponges, 42 species, 100 specimens. *Sponges,* . 100

Acalephs,
Polyps, } 446 species, 1500 specimens.
Echinoderms, *Radiates,* . 1500

Mollusks, in alcohol, 500 species, 1000 specimens.
Shells, 4152 species, 8000 specimens.
Mollusks, . 9000

Worms, 110 species, 200 specimens.
Crustaceans, 150 " dry, 330 species in alcohol, about 1300 specimens.
Insects, 21000 specimens pinned, 5000 specimens in alcohol, of these 2000 species of the pinned have been catalogued.
Nests, 15 species, *Articulates,* . 27515

Fishes, 1000 species, 2000 specimens in alcohol, and about 200 specimens dry and mounted.
Reptiles, 400 species, 1000 specimens, principally alcoholic. (A fine collection of Turtles mounted.)
Birds, 100 species, 150 specimens in alcohol; 411 species, 500 specimens mounted.

Birds' nests,	50 species,	80 specimens.			
Birds' eggs,	180 "	425 "			
Mammals,	51 "	75 "	in alcohol;		
"	65 "	70 "	mounted;		
"	9 "	10 "	as skins.		
			Vertebrates	.	4510

Skulls of Mammals, 172 species, 230 specimens, of which 39 are human.

Skulls of Birds, . .	150 species,	200 specimens.		
Skulls of Reptiles, . .	27 "	27 "		
Skulls of Fishes, . .	10 "	10 "		
Skeletons of Mammals, .	12 "	12 "		
Skeletons of Birds, . .	4 "	5 "		
Skeletons of Reptiles, .	6 "	6 "		
Skeletons of Batrachians, .	10 "	30 "		
Skeletons of Fishes, .	8 "	8 "		
Parts of Skeletons, of Mammals,	8 "	8 "		
" " *Birds,*	10 "	10 "		
Teeth of Mammals, .	14 "	30 "		
Jaws of Fishes, .	15 "	20 "		
Horns and Antlers, .	43 "	43 "		
			Osteological collection	639

May not the Institute hope that its friends and the patrons of science will soon give that aid, which is so essential to promote its objects and to continue with success its usefulness in diffusing a knowledge of the works of the Creator and of the History of Mankind?

If an addition to our present accommodations and means could be obtained and a number of professional Naturalists, having the charge of the various branches of the department of Natural History, and also several assistants in the Library, who, in addition to the ordinary duties could classify and arrange the manuscripts, pamphlets, newspapers and other materials that appertain to the Historical department, be permanently attached to the institution, much good would be done to the cause of education in our community by well arranged collections and libraries, and also by free lectures illustrating the various objects of the Institute. Much could also be accomplished through the medium of our publications in advancing the cause of science, and also of historic research by the continuation of the printing of abstracts of wills, deeds and other documents which are deposited in the offices of the county of Essex, and other materials of an historical nature that may be obtained from various sources.

Letters were read from:

Prof. S. S. Parvin, Iowa City; Asst. Surg. B. G. Wilder, 55th Mass., Vol.; W. P. Alcott, Andover; Dr. Wm. Prescott, Concord, N. H.; J. W. P. Jenks, Middleborough; John Jenkins, Monroe, N. Y.; Rev. Joseph Blake, Gilmantown, N. H.; Edwin Harrison, St. Louis, Mo.; Dr. Simeon Shurtleff, Weatogue, Ct.; W. W. Jefferis, Westchester, Pa.; W. J. Beal, Cambridge; L. E. Chittenden, New York; Wm. H. Edwards, Newburgh, N. Y.; Wm. F. Hall, Boston; Theodore Howland, Sect. Buffalo Soc. Nat. Science; E. Lewis, Jr., Brooklyn, N. Y.; Chas. N. Hoyt, Providence, R. I.; John C. Trantivine, Philadelphia; Frederic Ware, Cambridge; Edward Norton, Farmington, Ct., relating to the publications: Elihu Hall, Athens, Ill.; Geo. C. Huntington, Kelley's Island, Ohio, relating to the collection of specimens: Prof. S. F. Baird, Smithsonian Institution, on business: F. C. Butman, accepting membership: A. G. Browne, department of the South, relating to the transmission of books for the library: Smithsonian Institution, acknowledging the receipt of publications: C. W. Felt, calling attention to Mr. Perkins' class in Phonography: A. Huntington, declining to be a candidate for the Presidency.

The following officers were elected for the ensuing year:

PRESIDENT.

FRANCIS PEABODY.

VICE PRESIDENTS.

Of Natural History—SAMUEL P. FOWLER. *Of History*—A. C. GOODELL JR.
Of Horticulture—J. F. ALLEN.

SECRETARY AND TREASURER.

HENRY WHEATLAND.

LIBRARIAN.

CHARLES DAVIS.

SUPERINTENDENT OF THE MUSEUM.

F. W. PUTNAM.

FINANCE COMMITTEE.

J. C. Lee, R. S. Rogers, H. M. Brooks, G. D. Phippen, Jas. Chamberlain.

LIBRARY COMMITTEE.

J. G. Waters, Alpheus Crosby, H. J. Cross, G. D. Wildes, William Sutton.

PUBLICATION COMMITTEE.

A. C. Goodell Jr., G. D. Phippen, Ira J. Patch, C. M. Tracy, Wm. P. Upham, R. S. Rantoul, F. W. Putnam.

LECTURE COMMITTEE.

Francis Peabody, A. C. Goodell Jr., G. D. Phippen, George Perkins, James Kimball, G. W. Briggs, F. W. Putnam.

FIELD MEETING COMMITTEE.

Geo. B. Loring, C. M. Tracy, S. Barden, S. P. Fowler, J. M. Ives, G. D. Wildes, E. N. Walton, Charles Davis.

CURATORS OF NATURAL HISTORY DEPARTMENT.

Geology—H. F. Shepard;
Mineralogy—C. H. Higbee;
Palæontology—H. F. King;
Botany—C. M. Tracy;
Comparative Anatomy—Henry Wheatland;
Vertebrata—F. W. Putnam;
Articulata—J. H. Emerton;
Mollusca—H. F. King;
Radiata—Caleb Cooke.

CURATORS OF HISTORICAL DEPARTMENT.

Ethnology.

William S. Messervy, M. A. Stickney, John Robinson.

Manuscripts.

W. P. Upham, H. M. Brooks, S. B. Buttrick, G. L. Streeter, G. D. Wildes, E. S. Waters.

Fine Arts.

Francis Peabody, J. G. Waters, J. A. Gillis.

CURATORS OF HORTICULTURAL DEPARTMENT.

Fruits and Vegetables.

J. M. Ives, J. S. Cabot, R. S. Rogers, John Bertram, G. B. Loring, S. A. Merrill, W. Maloon, A. Lackey, G. F. Brown, C. H. Norris, C. H. Higbee.

Flowers.

Francis Putnam, William Mack, Benj. A. West, Geo. D. Glover.

Voted; That the meetings on the first and third Mondays of each month be held at 4 o'clock P. M. until otherwise directed.

Voted; That the Curators of Horticulture be authorized to hold exhibitions of Fruits, Flowers, Vegetables &c., at such times and places as may be desirable; also to offer premiums and gratuities for specimens exhibited, under such regulations as they may adopt.

Voted; That Messrs. Goodell, Rantoul and Upham be a committee to prepare suitable resolutions expressive of the thanks of the Institute due to A. Huntington, the retiring President, for his valuable services during the four years which he has presided over the Institute.

On motion of Mr. Putnam, Chapter I, Section VI, lines five and six, of the By Laws were so amended as to read "make such use of the duplicates as may be beneficial to science."

George M. White, James A. Chamberlain, Jonathan Ropes and George Fowler, all of Salem, were elected Resident Members.

MONDAY, MAY 15. Regular meeting.

Vice President Goodell in the chair.

R. S. Rantoul, for the committee appointed at the annual meeting, submitted the following report which was adopted and a copy of it ordered to be transmitted to the retiring President:—

Whereas the Honorable Asahel Huntington having declined reelection to the Presidency of the Essex Institute after four years of acceptable service in that capacity, during which the Institute has prospered beyond precedent.

Therefore, *Resolved:* That we cannot forego this opportunity of putting upon record our appreciation of the virtues of his private character, and of the usefulness of his long professional and public career; together with the hope that he may hereafter look back upon his efforts, while President of this body, in behalf of sound learning, the diffusion of useful knowledge and the generous culture of letters, science and the arts, as not the least among the honorable services of a well spent life.

John L. Marks, William H. Silsbee, and Henry R. Gardner, of Salem, were elected Resident Members.

TUESDAY, JUNE 6. Adjourned Regular meeting.

Henry F. King in the chair.

On motion of the Superintendent it was *Voted:* That the thanks of the Institute be tendered to George C. Huntington, Esq., of Kelley's Island, Ohio, for the donation of a valuable collection of Fishes from Lake Erie, and also for his kindness in defraying the necessary expenses attending the same.

WEDNESDAY, JUNE 7. Field meeting at Nahant.

The society opened their series of Field meetings this day by a visit to the ever delightful retreats of Nahant. The number in attendance reaching over two hundred who took the regular conveyances from the Central Station, besides many who took other means and different hours for the passage. Arriving at the Methodist Chapel, which had been selected as the place of meeting, the company deposited their various provisions, and under the guidance of John Q. Hammond, Esq., the greater number set out to examine the curiosities of the place. Some, in search of particular objects, scattered here and there, to fish, or gather plants, or break the rocks for specimens of minerals. But most of the party made a circuit round the shore, passing the summer residences of Gen. Fremont, Prof. Longfellow, and that formerly of Prescott, the historian, as well as many more. The "Swallow's Cave" received due attention, as also those features of the eastern extremity, "Pulpit Rock," "Natural Bridge," "Castle Rock," and the "Spouting

Horn." A brief stay was made about the crumbling ruins of the old Nahant Hotel, whose scorched and shivered stones yet bear up against the elements. Some pursued their walk to the "Maolis Gardens," so fancifully laid out and bedecked by the late Frederick Tudor, and puzzled themselves with his unexplained devices. On returning, and after a plentiful repast, the party assembled in the Chapel for the formal exercises of the afternoon, when the meeting was called to order by A. C. Goodell, Historical Vice President.

Letters were read from:

Rev. E. C. Bolles, Portland, Me.; Trübner & Co., London, Eng.; Mrs. B. F. Mudge, Quindaro, Kansas; E. L. Layard, Director of the South African Museum, Cape Town; Elihu Hall, Athens, Ill.; C. M. Tracy, Lynn; Smithsonian Institution; J. J. Babson, Gloucester; G. B. Loring; A. Lackey, Marblehead; John Howarth, Boston; H. R. Stiles, New York; David Choate, Essex; Dr. Wm. Prescott, Concord, N. H.; Dr. T. M. Brewer, Boston; Chas. M. Wheatley, Phœnixville, Pa.; Prof. Richard Owen, Indiana State University; B. O. Peirce, Beverly; A. S. Peabody, Cape Town, Africa; Geo. C. Huntington, Kelley's Isl., Ohio; Prof. A. E. Verrill, Yale College; Prof. James Hall, Albany, N. Y.; W. W. Denslow, Inwood, N. Y.; W. W. Stewart, Custodian, Buffalo Soc. of Nat. Sciences; Brown, Brothers & Co,, Boston, on various business matters: Sidney S. Lyon, Jeffersonville, Ind.; Rev. Samuel Lockwood, Keyport, N. J.; Dr. T. M. Logan, Sacramento, Cal.; Thure Kumlien, Busseyville, Wis.; D. G. Tompson, Montpelier, Vt.; Prof. S. F. Baird, Smithsonian Institution; Prof. C. C. Hamlin, Waterville College; H. B. Lord, Ludlowville, N. Y.; Codman & Shurtleff, Boston; W. A. Haines, New York; E. S. Morse, Gorham, Me.; W. J. Beal, Cambridge; Dr. Frederick Brendel, Peoria, Ill.; Prof. S. N. Norwood, Columbia, Mo.; Rev. E. B. Eddy, Waltham; Dr. John G. Thomas, Riviére-du-Loup-en-bas, Canada; Prof. Alex. Winchell, Ann Arbor, Mich.; G. A. Boardman, Milltown, Me.; A. L. Russell, Quebec, Canada; Henry Bannister, Evanston, Ill., relating to the Publications: Maine Historical Society, acknowledging receipt of Publications; Albert G. Browne; J. Vincent Browne; F. Cox; transmission of Specimens: Prof. James Hall, Albany; Henry R. Gardner; accepting Membership: F. Peabody, accepting the office of President: G. B. Loring, accepting the office of Chairman of Field Meeting Committee.

Donations to the Museum and Library were announced.

Dr. George B. Loring being called upon said that he had been surprised in several ways to day. He had found evidences of greater antiquity on Nahant than he had supposed were to be seen on the continent. He had found a ruin here, whose worn and tottering stones showed more of the ravages of time than the broken arches of the Forum. He had been through Rome and many places in Europe of the older sort, but nowhere had he seen such worn and plainly antiquated piles as appeared on Nahant to day. Further on, he had found the very rocks carved with inscriptions in forgotten tongues. No one here to day could read them, no one could say what meant the '*epinon ek tes petras*' that still endured in the monumental granite of Nahant. Statues were here, but beyond the design of the old masters; frescoes, but wholly pre-Raphælite in their execution. He was full of wonder at what he had seen. He further spoke of the place as formerly a field for the simple agriculture of the early time, when the unambitious farmers drove their flocks here to graze by day and brought them home at night. He closed by an eloquent allusion to the restoration of peace, under which blessing the Institute could come to such delightful spots as this and continue its Field meetings.

John Q. Hammond, Esq., of Nahant, would speak in behalf of his townsmen and extend their welcome to the Institute in their visit to day. For himself, he felt little of the enthusiasm in the study of nature that some exhibited, but he could appreciate the purposes and the utility of the Society, and was glad to lend what help he might to promote its interests. A good field was surely here for exploration; students were constantly resorting to it, and its rare and curious wealth seemed only partly yet discovered.

C. M. Tracy, of Lynn, gave a brief account of some points in the botany of the peninsula. A remarkable fact appears in the almost total absence of all heath-like plants from this place. It is said there are a few huckleberry bushes here; but not a pyrola, laurel, lambkill, blueberry, swamp-pink, or any such thing could he ever find. On the contrary, the field chickweed, a most lovely plant for the lawn, grows here abundantly, though rare or wanting in all the region round. Formerly, it is said, Nahant had heavy forests; but the settlers destroyed them, and since then a tree can scarcely be made to endure the climate. Persevering care has, however, partly retrieved the error, and the place is growing far greener than for years before.

F. W. Putnam, of Salem, gave some description of the zoölogical specimens taken during the day.

Wm. J. Beals, of Union Springs, N. Y., said he had been born and brought up in a country where there were no rocks, no ocean, no evergreen trees, and he had heard of these things in his childhood, as the inventions in a pleasant story. A few years ago, he had made a pilgrimage to New England that he might see these things; and he had set himself down by the sea for half a day at a time, full of delight as a child. You who live in the midst of these things have no idea of their true beauty. He gave a description of the curious plant called the sundew, which traps insects by the adhesive drops on its leaves.

Caleb Cooke, late of Zanzibar, East Africa, excused himself from speaking on account of feeble health. He had enjoyed the day and its rambles, and at another time would be glad to speak.

Abner H. Davis and Emery S. Johnson, of Salem, were elected Resident Members. John da Costa Soares, of Mozambique, E. C. A.; George C. Huntington, of Kelley's

18

Island, Ohio; Professor Richard Owen, of New Harmony, Ind., and Professor Leo Lesquereux, of Columbus, Ohio, were elected Corresponding Members.

On motion of Mr. Tracy the thanks of the Institute were voted to the Proprietors of the Methodist Church for the use of their House for the meeting, and to Messrs John Q. Hammond, Edmund Johnson, George A. Perkins and other friends in Nahant for their favors on this occasion.

MONDAY, JUNE 14. Regular Meeting.

The President, Francis Peabody, in the chair.

Letters were read from:

W. W. Denslow, Inwood, N. Y.; S. M. Buck, Hancock, Mich.; A. S. Taylor, Santa Barbara, Cal.; Prof. J. G. Norwood, Columbia, Mo.; E. S. Morse, Gorham, Me.; Prof. Joseph Henry, Sect. Smithsonian Institution; N. Brown, Boston; Geo. C. Huntington, Kelley's Island, Ohio; H. W. S. Cleveland, Danvers; Andrew Lackey, Marblehead; Leverett Saltonstall, Newton, relating to general business and the transmission of books and specimens: Geo. Scarborough, Sumner, Kansas; Rev. A. P. Chute, Sharon, relating to the publications: J. J. H. Gregory, Marblehead; C. M. Tracy, Lynn; G. W. Skinner, New Bedford; N. E. Atwood, Provincetown, relating to the Field meetings: Abner H. Davis, accepting membership: G. H. Peirson, containing an invitation to join in the celebration of the fourth of July in Salem.

The Secretary submitted the following resolutions which were unanimously adopted.

The Institute having been invited to join in the celebration of the fourth of July in Salem:—

Resolved: That the coming anniversary of American Independence marks an epoch in the progress of civilization and the life of Nations to which no studious observer can be indifferent, and that this Historical Body cordially unite, in spirit, with the people of this community, in such demonstrations as shall impress upon the minds of all the character of the crisis through which the Nation has passed and the honorable place in history which awaits the defenders of their Country.

Resolved: That, whereas, the Members are widely scattered, and many of them will take part in the demonstration in connection with other bodies, the Institute, grateful for the polite invitation, deem it unadvisable to take any prominent position; at the same time they will cheerfully render such assistance in their power, consistent with their regulations, as may contribute to the interest of the occasion.

The President, as chairman of the committee on the "First Church," presented a final report of the doings of the committee. The frame of the Church has been removed and placed in the rear of Plummer Hall, encased in an external structure of suitable strength to which it is bolted and is seen projecting from the plastering on the inside of the building. The Committee, in giving the key of the building to the Institute, do so, with the sincere wish, that the Holy House may be preserved to those who come after us, and handed down, from generation to generation, as a valued trust.

Rev. G. D. Wildes, after some appropriate remarks, moved that the report be accepted and that the thanks of the Institute be tendered to the committee, for the faithful and successful performance of their duty, which was unanimously adopted.. [The report of the Committee will be published in full in the Historical Collections.]

R. S. Rantoul read a memoir of Major Thompson Maxwell, a soldier in the old French war, the Revolution, and the war of 1812.

On motion of Mr. Putnam, the memoir read by Mr. Rantoul was referred to the publication committee, for publication in the Historical Collections.

The thanks of the Institute were voted to Colonel E. F. Miller for the presentation of the original documents, from which the memoir of Major Maxwell had been compiled.

Mr. Putnam communicated, by title, a paper by E. S. Morse, on the "Classification of the Mollusca on the Principle of Cephalization," accompanied with a plate. Referred to the Publication Committee.

Donations to the Library and Museum were announced.

H. R. Lovett of Beverly was elected a Resident Member.

THURSDAY, JUNE 24. Field meeting at Standley's Grove in Beverly.

This second Field meeting of the season was attended by a large party from Salem and the adjoining towns, who, after a pleasant forenoon's ramble in the woods and over the Laurel Ground, and a collation in the grove, assembled on the platform, when the meeting was called to order by Dr. G. B. Loring, Chairman of the Field meeting committee, who made a few remarks upon the history of the town, alluding to several of the worthy persons who once lived within its limits, as the Cabots, Nathan Dane, President Willard, of Harvard University, and others.

After the reading of the records of the last meeting, letters were read from the following:

Thomas Bland, New York; Thomas Mc Ilwraith, Hamilton, C. W.; Capt. Alpheus Hyatt, Baltimore, Md.; Thomas M. Peters, Moulton, Ala.; John Bolton, Portsmouth, Ohio; S. I. Smith, Norway, Me.; A. L. Babcock, Sherborn, Mass.; E. S. Morse, Gorham, Me.; Thomas Meehan, Germantown, Pa., relating to the publications: Prof. Theo. Gill, Smithsonian Institution, applying for the use of specimens: Prof. Leo Lesquereux, Columbus, Ohio, accepting Membership and notice of transmission of books for the library: Asst. Surg. A. S. Packard, Camp near Washington; C. M. Tracy, Lynn; Dr. W. Prescott, Concord, N. H.; Rev. S. Barden, Rockport, on business: B. F. Mudge, Quindaro, Kansas, stating that he will send fossils to the Institute: W. H. Dall & R. E. C. Stearns, San Francisco, relating to an exchange of specimens.

Donations to the Library an l Museum were announced.

Rev. G. D. Wildes read a poem written for the occasion by Mrs. P. A. Hanaford.

Dr. Henry C. Perkins, of Newburyport, read a paper of which the following is an abstract.

Attempt to explain the formation, or development, of the cumulus or thunder cloud on the principles laid down by Mr. Espy in his Philosophy of Storms, modified somewhat as to the principle of the ascent of the air in the tornado and water spout.

On the 9th of August, 1852, a large cumulus cloud was observed in process of development, the cloud was soon capped by a dense, white vapor, (as if by a veil), showing that the air above the cloud was being lifted bodily (as it were) by the ascending column of hot air, above the dew point.

The dry bulb thermometer stood at 79° F., wet bulb at 69° F., the dew point being at 65° F.

The rain was soon seen to fall and in a few moments the lightning was observed, followed, at an interval of 45 seconds, by a clap of thunder. A heavy shower from this cloud fell at Hampton Falls.

Reckoning a fall in temperature of 1° for every 100 yards, the base of the cloud was about 1,400 yards above the surface of the Earth, and the top of the cloud about three times as high, or 4,200 yards.

If we suppose with Mr. Espy that the air cools in ascending at the rate of 1.5° for every hundred yards, the thermometer in and outside of the cloud would indicate a fall of 60° in the temperature in ascending 4,000 yards, unless warmed up by the condensation of the vapors in the cloud.

By the above ratio of descent of temperature, when the air has risen 1,700 yards, the temperature will have fallen between 25 and 26°, and in so doing, will, (according to Dalton's tables) after making due allowance for the increased space it occupies in ascending three miles, viz.: one-third, condense forty-four one hundred and fifteenths nearly of its vapor; which would be sufficient to heat up the air in the cloud 35°.

The expansion of the air in the cloud by the giving out of this amount of latent heat, viz.: 35,° would equal thirty-five four hundred and eightieths or one-fourteenth nearly, of its bulk, or of the space occupied by the ascending column of hot air.

Supposing now the base of the cloud to be at one mile above the earth, where the barometer may be taken as standing at 30 inches and the density of the air at unity, or one; at four thousand two hundred yards, its density would be .630, or one third less: or in other words the barometer would stand at 20 inches at or just above the cloud, and at 30 inches at its base: one-fourteenth of this difference would equal .71 of an inch of barometric pressure, which would express the fall of the barometer from expansion by the heat given out by the condensation of the vapors if the expansion was all in an upward direction.

On the supposition that the column of heated air or the cloud ascends to a point where the barometer would stand at 20 inches, the amount of rain which would fall would be about 1.6 inch, supposing all the vapors to be condensed and to fall on an area equal to the base of the cloud; and it would occupy about 30 minutes in falling: for when the dew point is at 65° the air contains about one seventy-seventh of its weight of vapor, and air at 80° dew point ascending on the principle of floating bodies, at the rate of 7 1-2 feet in one second, at 65°, would rise one tenth less rapidly, or at about the rate of 6 feet and 9 inches in a second.

Without doubt the Sun may and does, in the day time, aid in the development of the cumulus cloud. We learn from Mr. Wise, the Aeronaut, that the air in the base of and on the sunny side of the cloud is much warmer than at other parts, and these clouds are seldom formed in the night; but we apprehend that the *electricity given off by the condensation of the vapors* is, in many of these clouds, especially those giving rise to the tempest or tornado and the water spout, the great expansive power in their development: indeed, on no other principles but that of the convective discharge of electricity can be explained the uplifting and removal to great distances of heavy bodies, the drying up of ponds, or the phenomena noticed in the subjoined account of the tornado which has so recently occurred in Wisconsin.

"An awful tornado nearly destroyed the village of Viroqua, Wisconsin, Thursday week. One hundred and seventeen persons were killed and wounded. A correspondent of the N. Y. World gives the following particulars:—

The southern part of the village. for a strip near eighty rods in width, was swept away. Where stood handsome white houses, neat barns, and out buildings, nothing now remains but ruins. Gardens, garden fences, orchards, grape vines, floral shubbery, well-curbs, buggies, wagons, cutters, &c., &c., were caught up, whirled, shaken, dashed to fragments, and the pieces taken for miles beyond. Never was work of destruction more rapid or complete. The track of the whirlwind is as if some mighty river had rushed over the course, leaving thousands of odd fragments strewn with liberal yet spiteful power.

Trees were torn up by the roots and thrown rods away. Roofs, sides, doors, floors, chimneys, underpinning, and furniture of houses were pounded together, broken into fragments and fairly sown over the land. Log chains were twisted apart, stoves and plow castings broken, ready for the smelter's furnace. Tree tops were loaded with clothing, bed-clothes, feather beds, carpets, chairs, harnesses, calves, sheep, dogs, cats, and poultry, dead or writhing on points of branches which had themselves been broken. Timbers have lodged in the tops of tall oaks, or, from their weight, borne saplings to the earth, and the saplings left covered with fragments of household goods as if hung out to dry. Doors, partitions, roofs, and floors of houses are found from five rods to three miles from where they belonged. Horses and cattle were killed or so badly maimed as to make their death an act of mercy. Fence rails, for ten years lying on the earth till imbedded therein, were whirled out. Stumps were torn up. Great rocks of twenty tons weight, were rolled, lifted, and broken by the mighty power.

Near the residence of John Gardner stands a tall oak rising about sixty feet from the ground. The wind whisked every leaf and small twig from the tree, leaving it looking as if dead. The house—a large white one—was taken so high in the air that it was seen above the tree tops, dashed to the ground, lifted again higher than before, whirled around and dashed roof down upon the earth a few rods from its foundation, and all but a few timbers borne away. Mrs. Gardner was in the house all the time; was spilled out in the second tumble and but slightly hurt, while an infant who was clinging fast in her arms escaped without a scratch or bruise!

In a school house were twenty-four children and a young lady teacher. The building was lifted high into the air, dashed upon the ground some distance from its foundation, again lifted about forty feet and dashed bottom up to the ground, and the fragments swept away. Eight children were killed and every other occupant badly injured. One little ten year old girl, whose thigh was lacerated and broken, when found in the fields begged the people to look for the others who were worse hurt than herself. The school house is not to be found.

Mr. Bennett was blown from his own demolished residence into a cellar near by, from which a house had been torn away. In a few seconds a little girl was thrown in by him for company. At the same time a horse was hurried in, striking Mr. Bennett and badly breaking a leg. The horse kicked and struggled to release himself from the rubbish which was "spilling" in upon the party, when Mr. Bennett tried to get a knife from his pocket that he might cut the poor animal's throat, and thus save the life of himself and the girl. At this moment a span of horses with part of their harness on were hurled in upon him and killed. The wagon to which they were attached went—the box to the west—the running gear into fragments and away over the field. The man who was in the wagon driving when the storm began was thrown like an arrow into an oak thicket thirty rods south from where he started, with fatal injuries."

Joseph H. Abbot, of Beverly, offered a few remarks, corroborative of the theory advanced by Dr. Perkins relating to the formation of thunder clouds, from personal observations.

Rev. Mr. Spaulding, of Newburyport, said a few words, expressive of his gratification at being present and of the increasing popularity of the study of Nature.

John I. Baker, of Beverly, welcomed the Institute, and thanked them for holding a meeting in the town.

C. M. Tracy, of Lynn, explained the plants which had been collected during the day, interspersing his remarks with many pleasant allusions, especially in respect to the Laurel found in such profusion, as though "to the Manor born," and thought that the reputation of Beverly could well "rest upon her laurels."

F. W. Putnam spoke of the nest of a Red-eyed Vireo, which he had found on a small oak in a swamp. The nest contained two eggs of the Cowbunting and none of the Vireo, having evidently been deserted as soon as the Cowbunting's eggs had been laid.

Joseph D. Tucke of Beverly presented a Lieutenant's commission given by Gov. Dudley of Massachusetts to Thomas Whittridge of Beverly, April 23, 1707.

R. S. Rantoul read a few extracts from the memoir of Thomas Maxwell, a Revolutionary hero.

Rev. G. D. Wildes offered some reminiscences of the brave young men who had achieved our National Independence.

The Secretary read the following communication:—

"C. M. Tracy, of Lynn, one of our esteemed Members and Curator of Botany, delivered on Saturday last the closing lecture of a course of eight on Botany. This course gave great satisfaction and was much admired by an appreciative audience. Before separating a meeting was called to order and Professor Crosby was invited to preside. James Upton, after a few appropriate remarks, introduced the following resolutions, which were unanimously adopted:—

Resolved; That we have listened with much satisfaction to the course of lectures by Mr. C. M. Tracy, of which the concluding one has been delivered this afternoon, and that the subject has been presented by him with a discrimination of thought and felicity of language as to demand some special token of our appreciation; we therefore tender to Mr. Tracy the thanks of this audience for his very successful efforts to interest us in his favorite study, the science of Botany.

It was then *Voted*, that a notice of these lectures with a copy of this resolution be communicated at a meeting of the Essex Institute with a request that the same be entered upon the records."

On motion of the Secretary it was *Voted:* That the above communication be entered upon the records.

George W. Pousland and J. Vincent Browne Jr., of Salem, were elected Resident Members.

On motion of C. M. Tracy it was *Voted:* That the thanks of the Institute be tendered to Charles Davis· Esq., Miss Sarah J. Tittle and other citizens of Beverly, for the kind interest they have manifested and the assistance they have afforded in carrying out this meeting.

Additions to the Museum and Library during April, May, and June, 1865.

TO THE NATURAL HISTORY DEPARTMENT.

BY DONATION.

ADAMS, SAMUEL, Hamilton. *Attacus cecropia* from Hamilton.

ALLEN, J. F., Salem. Larvæ and Imago of Lepidopterous Insects from the Grape vine.

BAKER, DAVID, Andover. Cast off skin of a Black Snake, 5 feet 2 inches in length.

BENNETT. MRS. A., Salem. Flying fish from Atlantic Ocean. Teeth of a Squid, Crustacean and Centipede from the East Indies.

BOLLES, REV. E. C., Portland, Me. Several Insects from Portland.

BROOKS, H. M., Salem. *Attacus polyphemus* from Salem.

BROWN JR., BENJ., Salem. *Papilio Turnus* from Salem.

Buffalo Society of Natural Sciences, Buffalo, N. Y. Collection of 180 species of Plants from the vicinity of Buffalo.

Buttrick, S. B., Salem. Several Minerals.

Carlen, Samuel, Salem. *Papilio Turnus* from Salem.

Carlton, Frazer, Salem. *Attacus cecropia* from Salem.

Colcord, Mrs. Helen M., South Danvers. Male, female, and young Red-winged Blackbird from South Danvers.

Conway, Mrs., Salem. Nuts from South America.

Cooke, Caleb, Salem. Collection of Plants from Zanzibar, Africa. *Ophidian* from a well in Zanzibar, Africa. Young Partridge from Salem. Parasites from the Cod, Haddock, and Skate, from Salem Harbor.

Crosby, Mrs. A., Salem. Cocoon of *Attacus cecropia* containing a grain of corn between the two layers of the cocoon.

Denslow, W. W., New York. 34 species of Plants, mostly collected by Audubon on his trip to the Rocky Mountains.

Eagleston, Capt. J. H., Salem. Sulphur from Isl. of Formosa.

Eaton, Peter E., South Danvers. Head of a Barbyroussa Hog killed on the Coast of New Zealand in 1845, by Charles H. Ingalls.

Emerton, James H., Salem. Gall flies and their parasites from Oak galls. Aquatic larvæ and larva of a Beetle from Swampscott.

Felt, John G., Salem. *Meloe angusticalis* from Salem.

Flint, C. H., Salem. *Bombus pennsylvanicus* from Salem.

Hammond, J. L., Salem. Musk gland of the Musk Deer from China.

Hathaway, Benj. F., Salem. Chicken with four legs.

Heath, N., Salem. Collection of Insects from Salem.

Hill, Benj. D., South Danvers. Barnacles from the Ship "Said Bin Sultan."

Hobart, Miss, Salem. Hymenopterous Insect and a Centipede from Honolulu, Sandwich Islands.

Hooper, Nath'l., Salem. Seeds from South America.

Huntington, Geo. C., Kelley's Island, Ohio. 25 species, 38 specimens of Fishes from Lake Erie.

Ives, J. M., Salem. Lepidopterous Insect from Salem.

Ives, J. S., Salem. Surf Duck from Salem Harbor.

Janes, Joseph P., Topsfield. *Attacus Luna* from Topsfield.

Jones, E. W., Salem. Collection of Insects and Spiders from Salem.

Kittredge, Miss, Beverly. *Attacus cecropia* from Beverly.

Mack, Miss Harriet O., Salem. 16 species of Shells.

Marks, J. L., Salem. Pudding stone from Long Island Sound. Breckenridge Coal from Kentucky.

Morgan, Miss Rebecca, Salem. Twig of a Clove tree with the leaves and fruit.

Nelson, S. A., Newburyport. Minerals from Dedham, Boxford, and Newburyport.

ORDWAY, COL. ALBERT, 24th Mass. Inf't. Specimen of *Madrepora* from Florida.

PERLEY, THOMAS W., Topsfield. *Attacus cecropia* from Topsfield.

PICKMAN, H. D., Salem. Several Fishes from Massachusetts Bay. Parasites from a Sculpin.

PORTER, E. J., Salem. 25 species of Insects from Salem.

POUSLAND, GEORGE, Salem. Skate from Salem Harbor.

PUTNAM, MRS. EBEN, Salem. Several Insects and Cells of the Mud wasp from Providence, R. I.

PUTNAM, E. L., Salem. 2 Cockroaches from Santa Cruz, British Honduras.

PUTNAM, F. W., Salem. Nest of the Red-eyed Vireo, containing two eggs of the Cow Bunting, from Beverly.

ROPES, T., Salem. Larvæ of an insect very destructive to the Tartarian Honeysuckle.

SAFFORD, JOSHUA, Salem. Malformed Claw of a Lobster.

SANBORN, FRANCIS G., Salem. Collection of Spiders from St. Louis, Mo.

SAUNDERS, MISS MARY, Salem. Eggs of a Lepidopterous Insect from an Apple tree.

SHEPARD, H. F., Salem. Hydroids and Polyzoa from Atlantic Ocean, Lat. 28°N., Long. 60°W. Collected April 1, 1865.

SHEPARD, S. A. D., Salem. Male and female Dragon flies from Salem.

SHREVE, CAPT. S. V., Salem. Malformed egg of a Hen.

SMITHSONIAN INSTITUTION, Washington, D. C. 66 species, 125 specimens of Bird's Eggs, principally from Arctic America.

SYMONDS, N. G., Salem. Embryo Skate taken from an egg found in Salem Harbor, May 1st, 1865.

TITTLE, MRS. S. J., Beverly. Spider from Beverly.

TUCK, J. D., Beverly. *Sesia pelasgus* from Beverly.

UPTON, WALTER, Salem. 2 young Hawks taken from a nest in Beverly.

VERRILL, PROF. A. E., New Haven, Ct. Collection of 46 Minerals from various localities.

WHEATLAND, DR. HENRY, Salem. Lepidopterous larva from the Birch.

WHITE, GEO. M., Salem. 3 species of Spiders from Salem.

YALE COLLEGE, New Haven, Ct. Dry specimens of *Astrophyton Agassizi*, *Asterias sp.*, *Solaster endica*, and *Solaster papposus* from Eastport, Me.; collected by Prof. Verrill, 1864. Collection of Minerals from various localities.

TO THE HISTORICAL DEPARTMENT.

BY DONATION.

ADAMS, SAMUEL, Hamilton. Two Stone Chisels of the Agawam Indians found in Hamilton.

BALCH, JOHN H., Newburyport. $5 bill on the Lincoln and Kenebec Bank, dated 1806.

BOWDITCH, MRS. REBECCA, Salem. Embroidered Mourning piece to the memory of Maj. Anthony Morse, 1803.

BROWNE, COL. ALBERT G.; 10 inch Shell invented by Capt. James Harding, Confederate Ordnance officer, made at Charleston Arsenal and used against Iron clad Ships. Large Shell fired from Battery Putnam. 10 inch shot from Charleston. Torpedo from Charleston Harbor. 50 specimens of Confederate local scrip of various denominations. 50 cents, 1, 2, 5, 10, 50, 100, 500, dollars Confederate States currency.

BROWN, J. VINCENT, Salem. Ballot thrown in the 8th Congressional District of Virginia, Nov. 6, 1861.

CARPENTER, D. B., U. S. Sanitary Commission. Canteen, Bullets and pieces of Shell from the Battlefield of Vicksburg.

CHAPEL, W. F., Salem. Relics from Gosport Navy Yard.

CHASE, GEO. H., Salem. Confederate Scrip of various denominations.

DENNIS, CAPT. JOHN, Beverly. Hand Grenade thrown from the fort at Port Hudson at the time of Gen. Banks' attack, July 8, 1863. Japanese Sword.

EAGLESTON, CAPT. J. H., Salem. Water Jar from Manila.

HAMMOND, J. L., Salem. Priest's Robe made by the Rebels while in Nankin. Brick from the Porcelain Tower of Nankin.

JOHNSON, W. B. F., Salem. War Club from Feejee Islands.

MARKS, J. L., Salem. Sword blade from the Plains of Abraham.

PUTNAM, F. W., Salem. $1 note of the Hungarian fund, dated New York, Feb. 2d, 1852.

RUST, JOHN O., Salem. 6 inch Cannon Ball from the Rebel Steamer Merrimac.

SHORT, MISS LYDIA ANN, Salem. Two Memorial Pitchers made during the Revolution. Chinese Umbrella.

SMITH, WARREN A., Chelsea. Two 10 cent Confederate Postage Stamps.

TRAILL, H. S., Marblehead. Pieces of the Confederate flag from Fort Pulaski.

UNDERWOOD, JOSEPH, Marblehead. 50 cent Confederate check on the Miss. Central R. R. Co.

VAUX, WM. S., Philadelphia. 14 Continental bills of Pennsylvania, Delaware and Maryland, of various denominations.

TO THE LIBRARY.

BY DONATION.

ANDREWS, CHARLES H. The New Whole Duty of Man, 1 vol., 8vo, London, 1788.

ANDREWS, MRS. J. H. Wheaton's Enquiry into the right of search, 1 vol., 8vo, Phil., 1842. Patent office Report, agric. 1857. 1 vol., 8vo, Washington, 1858.

BOSTON, CITY OF. Boston city Documents for 1864, 2vols., 8vo.

BROOKLYN MERCANTILE LIBRARY ASSOCIATION. 7th Annual Report, March 30, 1865, 8vo, pamph.

BROOKS, HENRY M. Statutes at Large passed 2d sess. 37th cong. U. S. A., 1861—2, 8vo, pamph.; Monthly Miscellany of Western India, 1 vol., 8vo, Bombay 1850; Youth's Primer by Jona. Fisher, 1 vol., 16mo, Boston, 1818; Cowper's Task, 1 vol., 16mo, Boston, 1819; Alden's Speaker, 1 vol., 12mo, Boston, 1810; Schoberl's Persia, 1 vol., 12mo, Philadelphia, 1828; Forbes' Maps of Richmond and its Fortifications.

BROWNE, ALBERT G. F. Hoffmanni Ommium Physicomed Supplementum, 1 vol., folio, Geneva, 1769; Harrington's Oceana, 1 vol., 8vo, Dublin, 1733.

BUFFALO'S YOUNG MEN'S ASSOCIATION. 29th Annual Report, 8vo, pamph.

BURGESS, GEORGE. Burgess' Address at the Funeral of Rt. Rev. T. C. Brownell, Jan. 17, 1865, 8vo, pamph.

CHASE, GEORGE C. Friends' Review, vol. xviii, Nos. 28 to 41, incl.

CHASE, GEORGE H. Richardson's Speech in convention of Virginia, April 4, 1861, 8vo, pamph., Richmond, 1862.

CHASE, MRS. GEORGE H. The Sanitary Reporter, three Numbers; The Sanitary Commission Bulletin, four Numbers; Pamphlets of Sanitary Commission, seven.

CLEVELAND, H. W. S., Danvers. Boston Courier from Jan. 1861, to June, 1865, 9 vols., folio; North American and United States Gazette from April, 1861, to June, 1865, folio, 5 vols., Philad.

COLCORD, MRS. H. M., South Danvers. The Pious Christian Instructed in the Catholic Church, 1 vol., 12mo, Dublin, n. d.

CURWEN, GEORGE R. Spirit of Missions, 34 Nos. The Church Almanac for 1864, 16mo, pamph. The Protestant Epis. Almanac for 1864, 16mo, pamph.

HALDEMAN, S. S., Columbia, Penn. Notes on Wilson's Readers, 12mo, pamph., Columbia, 1864.

HAMMOND, J. L. Legge's Chinese Classics, vols. 1 and 2, 8vo, Hongkong, 1861.

Hanaford, Mrs. P. A., Reading. The Martyred President by Mrs. P. A. H., 8vo, pamph. Pamphlets, twenty.

Holden, N. J. Mass. Legis. Documents for 1865, 4 vols., 8vo.

How, Henry, King's College, Windsor, N. S. On Magnesia Alum by Prof. How, 8vo, pamph. On Economic Geology of Nova Scotia, pt. 1, 8vo, pamph. On Mordenite, 8vo, pamph. On the Waters of the Mineral Springs of Wilmot, 8vo, pamph.

Hunt, Thomas. A School Book in Chinese, 1 vol., 8vo. A Visiting card of Howqua.

Johnson, Mrs. Lucy P. Monthly Journal of Amer. Unit. Association, vols. 4 and 5, 12mo, Boston, 1863—4. Fowler's English Grammar, 1 vol., 12mo, New York, 1860; The New Testament, 1 vol., 12mo, New York, 1819; Pierpont's National Reader, 1 vol., 12mo, Philad., 1854.

Kimball, Miss Elizabeth. Liberator for 1864, 1 vol., folio, Boston.

Kimball, James. Proceedings of the Supreme Council of the Northern Jurisdiction U. S. A., 8vo, pamph., Boston, 1864; Proceedings of Gr. R. A., Chapter of Mass. for 1864, 8vo, pamph.

King, Henry F. Scientific American, vols. xi, xii, xiii, xiv. New Series, vols. i, ii, iii, iv.; Blackwood's Edin. Magazine, 29 Nos.; Littell's Living age, 29 Nos.

Lapham, J. A., Milwaukie, Wis. Lapham's Maps of Wisconsin, showing the remarkable effect of the Lake in elevating the mean temperature of Jan'y., and depressing that of July, 1865.

Layard, E. L., Cape Town, S. Africa. Catalogue of the South African Museum, pt. 1, 16mo, pamph.

Lee, Miss Harriet R. Collection of Hand Bills and Programmes.

Lee, Higginson & Co., Boston. Rand & Avery's Specimens of Printing, 1 vol., 8vo, Boston, 1865.

Lewis, Winslow, Boston. A description of the City Hospital of Boston, 8vo, pamph.

Logan, T. M., Sacramento, Cal. Reports of Bd. of Agric. to California State Agric. Soc'y., Jan'y. 26, 1865, 8vo, pamph.

Lord, N. J. Files of Boston Post for Jan'y., Feb. and March, 1865.

Loring, George B. Boston Post for March, April and May, 1865.

Lowe, Charles, Somerville. Lowe's Sermon on the death of Lincoln at Charleston, April 23, 1865, 8vo, pamph., Boston, 1865; Lowe's Discourse, June 4, 1865, on the condition and prospects of the South, 8vo, pamph, Boston, 1865.

Mack, Miss Harriet O. "Boatswains Whistle," pub. at National Sailors' Fair in Boston, November, 1864.

Matthew, George F. Cupriferous Rocks of South Eastern New Brunswick, 8vo, pamph., St. John, N. B., 1865.

NORWOOD, J. G., Columbia, Mo. Norwood's Notice of Producti &c., found in Western States, 1 vol., 4to; Norwood's Illinois Geological Survey, 8vo, pamph.; Several Pamphlets relating to University of Missouri.

ODELL, CHARLES. Eliot's Biographical Dictionary, 1 vol., 8vo, Salem, 1809.

OWEN, RICHARD, Bloomington, Ind. Owen's Key to the Geology of the Globe, 1 vol., 8vo, Nashville, 1857; Owen's 2d Report on Geology of Arkansas, 1 vol., 8vo, Phil., 1860; Owen's Report on the Geology of Indiana, 1 vol., 8vo, Indianapolis, 1862; Catalogues of Indiana University, 1863—4, 1864—5, 8vo, pamph.

PATTERSON, ROBERT, Pennsylvania. A Narrative of the Campaign in the valley of the Shenandoah, by R. Patterson, 1 vol., 8vo, Phil., 1865.

PEASE, GEORGE W. 3d Ann. Rep. U. S. Christian Commission, Jan. 1, 1865, 8vo, pamph.

PERLEY, JONATHAN. Hepworth's Address at the Starr King Lodge of Masons, April 17, 1865, 8vo, pamph.

PICKMAN, WILLIAM D. American Museum, vols. 2, 3, 6, 7, 9, 10, 11, 12. 8 vols., 8vo; Gray's How Plants Grow, 1 vol., 12mo, New York, 1859; Whelpley's Compend of History, 1 vol., 12mo, Boston, 1821; Goodrich's Early History of Virginia &c., 1 vol., 16mo, Boston, 1833; Wakefield's Botany, 1 vol., 12mo, Phil., 1818; Guide to the Lakes of Cumberland &c., 1 vol., 16mo; Ronna's Dictionnaire Francais—Italien, 1 vol., 16mo; Paris, n. d.; Spirit of the Annuals for 1830, 1 vol., 16mo, Phil., 1830; Willis' Legendary, vol. 1, 1 vol., 12mo, Boston, 1828; The Mysteries of Udolpho, 3 vols., 12mo, Phil., 1800; Goodrich's United States, 1 vol., 16mo, Boston, 1827; Memoirs of Sherburne, 1 vol., 12mo, Providence, 1831; Rambles in Italy in 1816—17, 1 vol., 8vo, Baltimore, 1818; Staunton's Embassy to China, 1 vol., 8vo, Phil., 1799; Life of Lafayette, 1 vol., 16mo, Boston, 1835; Memoir of James Jackson Jr., 1 vol., 16mo, Boston, 1836; Maitland's Narrative of the Surrender of Napoleon, 1 vol., 12mo, Boston, 1826; China and the English, 1 vol., 16mo, New York, 1835; Cathrall's Buchan, 1 vol., 8vo, Phil., 1799; Angleskay Grammatika, 1 vol., 8vo; The Children's Week, 1 vol., 12mo, Boston, 1830; Das Newe Testament Nach D. Martin Luthers, 1 vol., 12mo, Frankfort, 1830; The Saracens, 2 vols., 12mo, New York, 1810; Minot's History of the Insurrection, 1 vol., 8vo, Boston, 1810; Ostrander's Mensuration, 1 vol., 8vo, New York, 1833; History of the Trial of Warren Hastings, 1 vol., 8vo, London, 1796; Joyce's Scientific Dialogues, 3 vols., 16mo, Phil., 1825; Comstock's Chemistry, 1 vol., 12mo, Hartford, 1831; Conversations on Chemistry, 1 vol., 12mo, New Haven, 1809; Blair's Grammar of Philosophy, 1 vol., 16mo, Hartford, 1822; Fowle's Linear Drawing, 1 vol., 12mo, Boston, 1825; De Porquet, the Turning of Eng.

Idioms into French, 1 vol., 12mo, Boston, 1833; De Porquet's Parisian Phraseology, 1 vol., 12mo, Boston, 1833; Comstock's Practical Elocution, 1 vol., 12mo, Phil., 1837; Dictionnaire des locutiones vicieuses, 1 vol., 16mo, Paris, 1813; The Traveller's Manual in Eng., Fr., Ger., and Ital., 1 vol., 16mo, Coblentz, 1847; The Holy Bible, 1 vol., 8vo, Boston, 1831; The Holy Bible, 1 vol., 16mo, Concord, 1838; Key to Murray's English Grammar, 1 vol., 12mo, Concord, 1820; Woodbridge's Geography, 1 vol., 16mo, Hartford, 1825; Adam's Latin Grammar, abridged, 1 vol., 16mo, New Haven, 1825; Putnam's Analytical Reader, 1 vol., 12mo, Dover, 1827; Sherwin's Algebra, 1 vol., 12mo, 7th ed., Boston, n. d.; Colburn's Key to Arithmetic, 1 vol., 12mo, Boston, 1829.

PUTNAM, MRS. E. A. 3 Pamphlets.

PUTNAM, F. W. and PACKARD, A. S. Putnam and Packard's Notes on the Humble Bees and their parasites, 8vo, pamph.

ROBERTS, WILLIAM S. Asiatic Journal, vol. 1, 8vo; Stewart's History of the Discovery of America, 1 vol., 8vo. Pamphlets, 3.

SAFFORD, MRS. JOSHUA. Washington's Farewell Address, 1 vol., 24mo, Newburyport, 1812.

SALISBURY, J. H.. Cleveland, Ohio. Catalogue of the Charity Hospital Medical College of Cleveland, Ohio, 8vo, pamph, Cleveland, 1865.

SHAEFER, P. W., Pottsville, Pa. Maps of the Anthracite Collieries and the Pottsville, Lehigh, Mahonoy and Shamokin Coal Basins in Penn., by P. W. Shaefer, 1859.

STIMPSON, WILLIAM, Chicago, Ill. Stimpson's Synopsis of Marine Invertebrata, 8vo, pamph.; Stimpson's Malacoölogical Notes, No. 1.

STONE, B. W. Memorandum in relation to the Gold Mines of the Chaudiere in Canada, 8vo, pamph.; Circular of the Dawn Petroleum Company.

TRASK, J. H., Wenham. Reports of Selectmen and School Committee of Wenham, March, 1865, 8vo, pamph.

TUCKER, W. P. Bishop Burgess' 5th Charge, July 9, 1862, 8vo, pamph.

WARD, CHARLES. Journal of Commerce Jr., New York; Essex Statesman, Salem.

WATERS, J. LINTON, Chicago, Ill. Receipts and Expenditures of Chicago, from April 1, 1864 to April 1, 1865, 8vo, pamph; A Guide to Illinois Central R. Road lands, 8vo, pamph., Chicago, 1865.

WHEATLAND, HENRY. The Olive and the Pine, 12mo, 1 vol., Boston, 1859; Record of an Obscure Man, 1 vol., 12mo, Boston, 1861; Success in Life, The Mechanic, by Mrs. L. C. Tuthill, 1 vol., 12mo, New York, 1860.

WILDES, J. H., San Francisco, Cal. 12th Annual Report of Mercantile Library Association of San Francisco, 8vo, pamph.; Views of the works of the Gould & Curry Silver Mining Company, Virginia City, N. T., 1 vol., 4to; Life of Joseph Vico, a Japanese who was rescued by the "Fennimore Cooper," an account of his travels in the United States, in the Japanese Language, 1 vol., 8vo.

WILLSON, E. B. Willson's Review of Ecclesiastical Proceedings in Brooklyn, Conn., 8vo, pamph., Worcester, 1818; Adam's Historical Discourse at Templeton, 8vo, pamph., Boston, 1857; Willard's Historical Discourse at Deerfield, 8vo, pamph., Greenfield, 1858.

YOUNG, S. J., Librarian of Bowdoin College. Catalogue of Bowdoin College, 1865, vo, pamph.

BY EXCHANGE.

AMERICAN ACADEMY OF ARTS AND SCIENCE. Proceedings, vol. ix., pages 341 to 364 incl.

AMERICAN ANTIQUARIAN SOCIETY. Proceedings of Special Meeting, Jan. 17, 1865, on the Death of E. Everett. 8vo, pamph.

AMERICAN PHILOSOPHICAL SOCIETY. Proc'd., vol. i, pages 1 to 48.

BOSTON SOCIETY OF NATURAL HISTORY. Proceedings, vol. ix., pages 305 to 375.

CANADIAN INSTITUTE. The Canadian Journal for March and May, 1865.

EDITORS. Savannah Daily Herald; North American Review; Historical Magazine; American Journal of Science; Florida Union; Home Evangelist; Salem Observer; Lynn Weekly Reporter; Essex Banner, (Haverhill); Haverhill Gazette; Lawrence American.

FIRELANDS HISTORICAL SOCIETY. The Firelands Pioneer, vol. vi., 8vo, Sandusky, 1865.

IOWA HISTORICAL SOCIETY. The Annals of Iowa for April, 1865.

MONTREAL SOCIETY OF NATURAL HISTORY. The Canadian Naturalist and Geologist for February, April and June, 1865.

MUSEUM OF COMPARATIVE ZOÖLOGY AT CAMBRIDGE. Annual Report of the Trustees for 1864, 8vo, pamph.

NEW ENGLAND HISTORIC-GENEALOGICAL SOCIETY. N. E. Hist. Gen. Register for Jan. and April, 1865.

PENNSYLVANIA HISTORICAL SOCIETY. Resolutions on President Lincoln, April 27, 1865, 8vo, pamph.

PHILADELPHIA ACADEMY OF NATURAL SCIENCES. Proceedings for Jan., Feb. and March, 1865.

TRÜBNER & Co., London. Trübner's Amer. and Oriental Literary Record, Nos. 1 and 2.

ZOÖLOGISCHE GESELLSCHAFT, Frankfurt a. m. Der Zoölogische Garten, vol. 5, Nos. 7 to 12 incl.

MONDAY, JULY 3. Regular meeting.

Henry F. King in the chair.

William H. Osgood and Joseph C. Foster, of Salem; Robert R. Endicott and George Roundy, of Beverly, were elected Resident Members.

THURSDAY, JULY 13. Field meeting at Reading.

The first Field meeting held by the Institute beyond the limits of Essex County took place in the town of Reading. The party from Salem leaving in the ten A. M. train and arriving at Reading at about eleven o'clock.

Reading is an attractive looking town, containing many hills and groves, among which are pleasant drives and walks. This town was many years ago a part of Lynn. It also included South and North Reading, which were afterwards set off in response to local requirements. On arrival, the company immediately repaired to the chapel of the old South Congregational Church, where the refreshment baskets were deposited and Vice President Goodell announced the programme for the day. As the time was limited, no very long rambles could be taken, and the few hours were passed in examining the garden of Mr. Amos Cummings situated on "Prospect Hill;" the nurseries of Mr. J. W. Manning; the old burial ground where many interesting and quaint epitaphs were to be seen; and by a trip to the pond and adjacent fields and groves.

About one o'clock the party again assembled in the Chapel, and, after partaking of refreshment, adjourned to the Church where the regular meeting was organized with

Vice President GOODELL in the chair.

The Rev. Wm. Barrows, pastor of the society in whose church the meeting was held, welcomed the Essex Institute to the town of Reading, alluding to the fact that this town was once a part of Lynn, and was then known as "Lynn Village," and, therefore, properly within the range of the researches of the Institute.

The Chairman responded, thanking the people of the town for the interest they had manifested this day and for their successful efforts to make the visit of the Institute a pleasant one.

Rev. W. W. Hayward, of South Reading, read an original hymn, written by a resident of Reading, which was sung by the choir of the church.

The records of the last meeting were read, and donations to the Library and Museum were announced.

Letters were announced from:

E. W. Blatchford, Chicago, Ill.; Chas. J. Sprague, Boston; Joseph N. Howe, Boston; Prof. C. F. Chandler, Columbia College; Thomas Barlow, Canastota, N. Y.; J. Kirkpatrick, Cleveland, Ohio; Prof. J. C. Holmes, Lynn; Dr. A. Kellogg, San Francisco, Cal.; J. J. Haagensen, St. Thomas, W. I.; Dr. Frederick Brendel, Peoria, Ill.; Geo. W. Peck, New York; Prof. A. Winchell, Ann Arbor, Mich.; Chas. Stodder, Boston; G. Hastings Grant, New York, relating to the publications: Dr. Wm. Stimpson, Corr. Sect., Chicago Acad. Nat. Science; Prof. Theo. Gill, Smithsonian Institution; Prof. S. F. Baird, Smithsonian Institution; E. A. Samuels, State Cabinet; E. S. Morse, Gorham, Me.; Rev. E. C. Bolles, Portland, Me.; Prof. J. G. Norwood, Missouri State University; Dr. A. S. Packard, Jr., Brunswick, Me.; Prof. P. A. Chadbourn, William's College; G. A. Boardman, Milltown, Me.; Dr. Daniel Clark, Flint, Mich.; Prof. D. S. Sheldon, Griswold College, Davenport, Iowa; Tryon Reakirt, Philadelphia; Henry L. Hotchkiss, New Haven, Conn.; Mrs. P. A. Hanaford, Reading; A. J. Archer, Salem, on business, and acknowledging the receipt of specimens: Prof. Richard Owen, New Harmony, Ind.; Prof. E. D. Cope, Haverford College, Pa., accepting membership: Maine Historical Society; Albany Institute, acknowledging receipt of Publications.

A communication on the Geology of Reading by Mr. L. B. Pillsbury of Hopkinton, formerly principal of the High School in Reading, was read by the chair.

John M. Ives, of Salem, spoke of Birds, particularly of the Swallows, describing the habits of the various species known in this vicinity. He also alluded to the habits of the Robin, Cow Bunting, Wren, Cherry Bird and Canada Goose, relating several curious anecdotes illustrative of the peculiarities of a number of the species.

Dr. G. B. Loring, in connection with Mr. Ives' remarks, spoke of the habits of the Eaves Swallows, a number of which had built their nests on his barn. Dr. Loring also claimed to be something of a Reading man, having once had charge of a school in that town, and related some amusing experiences connected with his professional duties. His compensation was $15 per month and "board round." He said that Rev. Dr. Flint, Hon. Amos Kendall and Rev. Cyrus Pierce had also been school teachers in the town. He related an anecdote of Mr. Kendall, who, while Post Master General under Jackson's Administration, had astonished some Reading politicians who desired a change of location in the town post office, by asking why the petition did not bear the signatures of certain leading men whom he named. "What!" said they, "do you know the name of every man in the United States?" The truth was Mr. Kendall remembered the names of some of the citizens who had been his friends, while a school teacher, at the age of sixteen.

F. W. Putnam, of Salem, made a few remarks on the geology of the town, called forth by Mr. Pillsbury's paper, and then proceeded to describe the few insects and fishes which had been collected during the morning.

C. P. Judd, of Reading, occupied a few moments, quite acceptably, with some interesting reminiscences of the early history of the town.

Ezra F. Newhall, of Salem, was elected a Resident Member.

On motion of Dr. Loring it was *Voted:* That the thanks of the Essex Institute be presented to the Proprietors of the old South Church of Reading, for the use of their house; and to the Rev. W. Barrows; the members of his society; and other friends in Reading, for their kind attention to the members of the Institute during the day.

After the singing of "America" by the choir and a benediction by the Pastor, the meeting adjourned in time for the cars for home, and all were well pleased with their visit to the town and the hospitality of its inhabitants.

At the depot, the signal master called the attention of a number of the members to a pair of Blue Birds which had built a nest in one of the signal balls, from which a piece of the canvas had been torn. These birds, after raising one brood of young, had made another nest, by the side of the first, in which they had laid the eggs for a second brood. The signal ball, in which the nests were made, was lowered and hoisted about fifty times a day. The birds flying out as soon as the ball commenced its descent, and, alighting upon the fence near by, would wait patiently for it to be hoisted again, when they would at once return to their nest.

MONDAY, JULY 17. Regular meeting.

Vice President Goodell in the chair.

William E. Doggett, of Swampscott and Sarah B. Endicott, of Salem, were elected Resident Members.

The Secretary stated that the portrait of Gov. John Leverett, which was sent, at the request of Leverett Saltonstall Esq., to Mr. Howarth of Boston, to be cleaned and restored, had been returned to the Institute in excellent condition, without cost to the Society.

On motion of Judge Waters, it was *Voted:* That a committee be appointed to tender to L. Saltonstall Esq., of Newton, the sincere thanks of the Essex Institute for this mark of his esteem and liberality in behalf of the Institute.

Messrs. J. G. Waters, H. Wheatland and S. B. Buttrick were appointed on the committee.

THURSDAY, JULY 27. Field meeting at Georgetown.

The Fourth Field meeting of the season was held this day at Georgetown.

Georgetown is not, distinctively, an old town. Its antiquity is entirely borrowed from the interesting town of Rowley, of which, like Boxford and Bradford, it was formerly a part, having enjoyed an independent existence among the family of towns only since the year 1838, and consequently being a younger sister of the towns last named. It was known, before the separation, as "New Rowley." The original post office box, not over three or four feet long, through which the New Rowley mail passed, is still in existence, and may be seen at the native wine establishment of Messrs. M. Carter & Son. It bears the following painted inscription: "New Rowley and Georgetown Post Office, established March 17, 1824; Benj. Little, P. M. First quarterly return, $7.32. Last quarterly return, June 1, 1845, $117.96. Whole amount collected, $5,373.63."

Georgetown appears to be one of the most active and spirited of our country towns, where attention is given not only to farming, but, also, a considerable share of the capital of its men of means is devoted to the manufacture of shoes, giving steady employment to many. The crops in the town look flourishing and bid fair to be plentiful. Apples will be scarce here as elsewhere in New England; but berries, cheapest of all fruits, abound.

On the arrival of the party a cordial welcome was tendered by O. B. Tenney Esq. Chairman of the Selectmen, who offered the hospitalities of the place, and called attention to the various points of interest in the town. Numerous vehicles were also in waiting to convey the party to the various objects of interest. Among the places visited by the several parties, were the "Old Lull House," which is situated about two and a half miles from the village on the Newburyport road. It is owned by Mr. Gorham D. Tenney, who is proprietor of the adjoining farm, which comprises two or three hundred acres. Mr. Tenney is the son of Capt. Gorham P. Tenney, whose wife was the daughter of Dudley Lull, whose name still imparts a designation to the old house. When, in 1690, the war was being conducted against the French in Canada, the Indians became troublesome in the Provinces, and on Oct. 23, 1692, this old house, which is in that part of the Byfield Parish included in the town of Georgetown, was the scene of a massacre of which an account may be found in Gage's History of Rowley. At that time it was occupied by a Mr. Goodrich who, with his wife and two daughters, while engaged in his family prayers, on Sabbath evening, were killed by the Indians. Another daughter, named Deborah, aged seven years, was taken captive, but was redeemed the next Spring, at the expense of the Province. She died in Beverly, as appears by the records of the First Church in that town, where the entry reads, "Buryed, March 28, 1774, Deborah Duty, aged 88, a widow." Those who were killed are said to have been buried in one grave a few rods to the east of the house. The exact spot, as located by tradition, was pointed out.

Mr. Tenney, the present owner, was very courteous and attentive to those who visited his place, and, besides proffering acceptable comforts, exhibited, at the farmhouse which he occupies, some good specimens of Indian relics, such as a pestle, gouge, axe, and arrow-heads; all having been found upon his farm, which was evidently an Indian resort in the olden time. He conducted the party through the old house, which is now very dilapidated and of course unoccupied. It has undoubtedly undergone some alterations since the day when Mr. Goodrich and his family were murdered, but the huge fire-places, clay-cemented chimneys, and broad and ponderous beams, betoken decided antiquity.

The famous octagonal barn of Mr. Samuel Littell was visited by a large party. This barn is said to be the largest in Essex County, being about eighty feet in diameter, each of the octagonal sides being about thirty-two feet. It has two floors in addition to the basement, and is so constructed with reference to the rising ground upon which it is built, that upon each floor there is an entrance from the ground. Situated upon a natural eminence, the view from the top of it is very extensive. Here may be seen Pentucket Pond, not far distant; Rock Pond, which flows into it; Haverhill, Groveland and Bradford; the ocean far away, and a vast extent of surrounding country, including distant eminences, among which several peaks of the White Mountains could be distinctly traced.

"Bald Pate," the highest hill in Essex County, being 392 feet above the level of the sea, and the "Ridges" were visited, as well as the Burial Grounds and other places of interest, among which was the Vineyard of Messrs. Carter & Son, who carry on a large manufactory of native wines.

The Mineral Point Mine on Atwood's Hill, was also visited. This mine yields an inexhaustible quantity of brown ochre, with which a large number of the houses in Georgetown and vicinity are painted, and which has been quite an article of export.

Another party made an excursion to the pond where a number of botanical and zoölogical specimens were collected.

Soon after one o'clock nearly all the parties had returned from their rambles, and assembled once more at the Town Hall, where, in addition to the refreshments brought by the company, the Georgetown people had liberally contributed to the entertainment. The hospitalities having been duly partaken of, the meeting for discussion was then held, commencing at half past two o'clock.

Dr. G. B. LORING, Chairman of the Field meeting Committee, in the chair.

The Chairman opened the meeting with remarks relating to Georgetown, of an historical character, alluding to its ecclesiastical controversies; some witchcraft experiences; and the character of some of its public men.

After reading the records of the last meeting and the lists of donations to the Library and Museum, letters were announced as received since the last meeting from:

Prof. J. Wyman, Cambridge; Prof. A. Winchell, Ann Arbor, Mich.; Frank Stratton, Natick, Mass.; T. A. Cheney, Havana, N. Y.; T. Bland, New York; E. S. Morse, Gorham, Me.; A. Hyatt, Baltimore, Md.; I. C. Martindale, Byberry, Pa.; Prof. L. Lesquereux, Columbus, Ohio; Prof. A. N. Prentiss, Lansing, Mich.; E. W. Blatchford, Chicago, Ill.; Thure Kumlien, Bussyville, Wis.; A. Agassiz, Cambridge, Mass.; B. P. Mann, Concord, Mass.; Noble, Brothers & Co., New York, relating to the publications: Redwood Library and Athenæum; Maine Historical Society, acknowledging the receipt of publications: Isaac M. Long, Salem; Col. Albert Ordway, Richmond, Va., transmitting specimens to the Museum: Prof. J. G. Norwood, Columbia, Mo.; Dr. T. A. Tellkampf, New York; Wm. Couper, Quebec, Canada; W. S. O. Brinckloe, Philadelphia, Pa.; Prof. Richard Owen, New Harmony, Ind.; Rand & Avery, Boston; Lowell Bleachery; New York State Library; Joshua P. Haskell, Marblehead, Mass.; J. R. Newhall, Lynn, Mass.; J. K. Wiggin, Boston; Lyceum of Natural History of New York, on business matters.

James P. Cooke and David P. Carpenter, of Salem, were elected Resident Members.

Mr. C. M. Tracy, of Lynn, was called upon by the chair, and responded in his usual happy and ready manner, giving some account of his observations during the day, and mentioned some of the principal plants collected. Among these were the Orchis, Buttonbush, Cardinal Flower,

Clethra, Asters, Golden Rods, and other varieties. In this connection some discussion arose in reference to the parasitical character of the Indian Pipe.

C. L. Flint, Esq., Secretary of the State Board of Agriculture, being called upon, made some general remarks favorable to scientific research and commending the objects of the Institute.

Rev. J. L. Sibley, Librarian of Harvard College, followed, speaking of the importance of preserving old pamphlets and papers, as having an important bearing, aside from any historical value, in settling questions involving the rights of property. He mentioned several instances which had come under his observation, and said the Institute was doing a valuable work in this connection, besides exerting an influence that was felt all over the country.

Capt. Alpheus Hyatt, of Baltimore, made some remarks with regard to the general sac like plan of the Animal Kingdom, defining the Radiata as radiated sacs, the Articulata as articulated sacs, the Mollusca as the simple typical sac and the Vertebrata as sacs divided internally into two cavities. Capt. Hyatt adduced specimens of *Paludicella* and *Fredericella*, found during the forenoon ramble, as proofs of the specialization of the sac among the Mollusca, and gave in detail their anatomical and physiological peculiarities.

Mr. F. W. Putnam, of Salem, gave a brief abstract of his day's observations, and enlarged upon the habits of the gall fly, specimens of which he produced at the meeting.

Capt. J. G. Barnes, while he made no claim to being a naturalist, said he had no doubt Georgetown could present much that was worthy the attention of a careful scientific observer. He said that during the past four years we have been making history very fast; and he looked with local pride upon what had been done in his state and town for the maintenance of the union of the States, and suggested that it was the duty of this historical society to gather all facts and memorials tending to elucidate the history of this period.

Mr. A. C. Goodell, of Salem, gave some curious facts in regard to the names of several towns in the vicinity, and closed his remarks by reading a poem written for the occasion by a Salem lady.

Richard Tenney, Esq., Postmaster of Georgetown, gave some facts in the history of the town, especially in relation to its incorporation as a distinct municipality.

Mr. John Preston, of Georgetown, presented a leaf from a magnolia planted by George Washington at Mount Vernon.

A resolution of thanks, offered by Mr. Walton and seconded by Mr. Upham, was passed to Messrs. O. B. Tenney and Sherman Nelson, Selectmen; Maj. Moses Tenney, Capt. Barnes, Lieut. Wildes; Messrs. Stephen Osgood, John Preston, John Bradstreet, Isaac Wilson, Edmund Bailey, Chas. Carter, Samuel Wadleigh, Geo. W. Boynton, Jos. Folsom, Richard Tenney, Geo. Harnden, Wm. Horner, Robert Coker, Gorham D. Tenney, and other citizens of Georgetown, for their liberal and successful efforts in making the meeting a pleasant and instructive one.

On returning to the depot, at the close of the meeting, Mr. W. S. Horner, the depot master, displayed a few of his many Indian relics.

WEDNESDAY, AUGUST 9. Quarterly meeting.

N. WESTON, JR., in the chair.

On motion of F. W. Putnam, it was *Voted:* That the following be added as a final clause to Chapter IV of the By-Laws.

"Every facility in the power of the Superintendent, consistent with the welfare of the specimens, shall be offered to persons visiting the Museum for the purpose of study and comparison."

D. B. Hagar, J. Leonard Hammond and Elizabeth A. Putnam, of Salem, were elected Resident Members.

FRIDAY, AUGUST 18. Field meeting at North Andover.

The fifth Field meeting of the season was held at North Andover this day. About three hundred persons arrived in the train from Salem and assembled at the "First Church" before separating into small parties to visit the special objects of interest to each.

The zoölogists sought the brooks and streams and found many specimens amply rewarding them for their search; the botanists the woods and meadows for various flowers; the antiquarians for the relics of olden time.

This township was first settled in 1634. In the same year the following order of the court was issued respecting the land in Andover:

"It is ordered that the land about Cochichewick shall be reserved for an inland plantation, and whosoever will go to inhabit there shall have three years immunity from all taxes, levies, public charges, and services whatever, military discipline only excepted."

The land is uneven, rising into large hills, affording fine and delightful prospects and scenery. Dr. Dwight, in his travels, some sixty or seventy years since, says of the North parish of Andover: "Its surface is elegantly undulating, and its soil in an eminent degree fertile. The meadows are numerous, large, and of the first quality. The groves, charmingly interspersed, are tall and thrifty. The landscape, everywhere varied, neat and cheerful, is also everywhere rich." Hither many come from the crowded city to enjoy the recreation of the country; and where can a better place be found than this well known Summer resort?

The Church was founded in 1645 and consequently is one of the oldest in the Conuty. For seventy years the desk was very acceptably filled by the two Barnards—the Rev. Thomas, and his son the Rev. John—and "during their ministry the people enjoyed a series of peace and improvement beyond what is common." The second Barnard died 14th of June, 1758, aged 68 years; he left two sons, both distinguished clergymen. One was the Rev. Thomas Barnard of Newbury, afterwards of the First Church in Salem, and father of the Rev. Thomas Barnard, D. D., first minister of the North Church in Salem; the other was Rev. Edward Barnard of Haverhill, whose portrait by Copley is in the possession of the Essex Institute.

The Great Pond, so called, is a fine, clear basin of water, containing obout 450 acres, and is well stocked with fish. The outlet, known as Cochichewick brook, furnishes the power of several woolen mills, some of which belong to the estate of the late Eben Sutton, Esq., of South Danvers. This was visited by many, and from the adjacent hills fine views of the Merrimack, the city of Lawrence, and other places, were enjoyed.

The afternoon session was called to order at three o'clock.

Dr. George B. Loring in the chair.

The records of the last meeting were read. Donations to the Library and Cabinets announced.

Letters were read from:

Rev. Samuel Lockwood, Keyport, N. J.; Edward A. Brigham, Boston; C. F. Austin, Closter, N. Y.; The Abbé Brunet, Quebec, Canada; E. S. Morse, Gorham, Me.; Prof. James Hall, Albany, N. Y.; James Lewis, Mohawk, N. Y.; Dr. Julius Homberger, New York; Henry White, New Haven, Conn.; J. H. Stickney, Baltimore, Md.; S. S. Parvin of the Iowa Historical Society; Dr. L. R. Stone, U. S. V., Harper's Ferry, Va.; Prof. Jonathan Pearson, Schenectady, N. Y., relating to the publications: Prof. Richard Owen, New Harmony, Ind.; W. H. Pease, Honolulu, S. I.; S. Jillson, Feltonville, Mass.; S. H. Scudder, Sect. Boston Soc. Nat. Hist.; Dr. A. S. Packard, Jr., Brunswick, Me.; Capt. Alpheus Hyatt, Boston; James R. Newhall, Lynn, Mass.; Mrs. P. A. Hanaford, Reading, Mass.; J. Prescott, Supt. Eastern R. R.; William Merritt, Supt. Boston & Maine R. R.; H. J. Cross, Salem, relating to the collection of specimens and business matters; Albany Institute; New York Historical Society, acknowledging the receipt of publications: Mrs. Sarah B. Endicott, accepting membership.

E. W. Buswell, of Malden; James Hill, Henry. P Hendrick, William Haskell, A. T. Mosman, of Beverly, and Martha G. Wheatland, of Salem, were elected Resident Members.

On the table were three beautiful and very finely executed paintings of flowers by Miss Eliza B. Davis, for several years a resident of Salem, a lady long and very favorably known to our citizens as zealously devoted to this beautiful art.

The Chair made some remarks upon the early history of Andover, alluding to several incidents connected with the first settlers and their immediate descendants. He spoke of the Phillips family, and paid a high eulogium to several of its members who have been great benefactors to the cause of education, in the liberal endowments to several seminaries of learning, which bear the honored name of Phillips. He also spoke of Stevens, the founder of one of the woolen mills on the Cochichewick stream, as one of the pioneers in this branch of our domestic industry. The old Franklin Academy was mentioned, incorporated in 1801, and which had been highly beneficial to the parish and to the youth who have enjoyed its advantages equally under the superintendence of Mr. Simeon Putnam, and of Mr. Cyrus Peirce, the experienced and faithful teacher, and the first teacher of a Normal School in this State. In this connection he alluded to the late Gen. I. I. Stevens, who fell fighting for the cause of his country in the recent rebellion, and who displayed in boyhood and youth the same intrepidity and courage which marked his later career.

The Chairman, after some additional remarks of a similar tenor, called upon Mr. John M. Ives, of Salem, who continued his observations upon the habits of many of our birds, which he had commenced at the meeting in Reading, a few weeks since, with especial reference to the migration of several species.

The Chairman stated that Andover had long been noted for its large trees, mentioning a large elm transplanted by Mr. Jonathan Frye in 1725, and called upon Mr. Goodell to give some account of what he had seen during the day.

Mr. A. C. Goodell, Jr., replied giving an interesting account of the large elm tree which he visited, and which measures, two feet from the ground, about thirty five feet in circumference. He then spoke of his ride around the Great Pond, above alluded to, and the view from some of the high hills, concluding by mentioning some interesting reminiscences of the early history of Andover. The land, including Andover, Lawrence, &c., was purchased of Cutshamache, the Sagamore of Massachusetts, for twenty six dollars, sixty four cents, and a coat. The town was incorporated in 1646 by the name of Andover, receiving that name from Andover in Hampshire, England, whence many of the settlers came.

Mr. F. W. Putnam spoke of the Striped Snake and other species which were found in this vicinity. Referring to the snake bite case in Lowell, Mr. Putnam said he had himself been bitten by the striped snake and had never experienced any ill effects, and he thought that the effects said to have followed the bite in the Lowell case were wholly due to fear, as there was no venomous fang in the striped snake. He then alluded to the fishes, giving some account of the minnows and pointed out the differences between these and those found in salt or brackish water.

Mr. E. G. Parker, of Groveland, asked some questions respecting the Tent Caterpillar, stating that from some happy but unaccountable cause all the caterpillars of this species, in this vicinity, had not had the strength to complete their cocoons, or had died soon after forming them. Considerable discussion upon the Tent Caterpillar and other injurious insects followed, participated in by Messrs. Parker, Ives, Putnam and the Chair.

The Secretary announced that Rev. Stillman Barden, a member of the Field Meeting Committee, and an ardent friend of the Institute, who had felt a great interest in, and had been a constant attendant upon these meetings, had died at Rockport since the last meeting; and upon his motion a committee was appointed to prepare suitable testimonials of respect to his memory and worth.

Mr. C. Davis, of Beverly, offered the following vote, which was unanimously adopted:

Voted; That the thanks of the Institute are due to the proprietors of the First Church in North Andover for the use of their house; to the members of "Cochichewick Engine Co., No. 2", for the use of their building; to Messrs. Moses T. Stevens, Otis Bailey, W. P. Phillips, John Bertram, James B. Curwen, Matthew Poor, I. O. Loring, Mrs. Nath'l Stevens, and other citizens and temporary residents of North Andover, for their kind attentions during the day.

The meeting then adjonrned.

WEDNESDAY, AUGUST 23. Adjourned Regular meeting.

Judge Waters in the chair.

Prof. J. G. Norwood, of Columbia, Mo., was elected a Corresponding member.

THURSDAY, SEPTEMBER 7. Field meeting at Newburyport.

The sixth and last Field Meeting of the season took place this day. About three hundred and seventy five persons attended, the larger portion proceeding over the Eastern Railroad to Newburyport, and thence down the Merrimack to Salisbury Point. The party was so large that, in addition to the passenger barge usually employed, it was found necessary to charter a schooner, both being taken in tow by a powerful little tug boat called the "Thurlow Weed". The trip down the river, some three miles, occupied about half an hour, and upon arriving at the Point, the barge and schooner were run directly upon the sand beach, and the company landed without any difficulty. Here nearly two hours were spent, and a few improved the time by inspecting Fort Nichols, and rambling over Salisbury Beach proper, which extends several miles along the ocean side, and is one of the finest beaches on the coast. The heat was so intense, however, that but a small number improved the opportunity. The fort mounts ten or twelve guns. The parapet already shows signs of disintegration, the effect, probably, of the severe drought, and of the sun's rays which concentrate upon the sand heaps with overpowering intensity. There were several tents in the vicinity, occupied by "camping out" parties from up the river.

Returning to Newburyport, the company partook of the usual picnic dinner in the City Hall, and afterwards had an opportunity to visit the many places of interest in the city, including the Horton Memorial Chapel, the Whitefield Church, the Copley Portraits, the Public Library, and other objects of note heretofore described.

At three o'clock, the afternoon meeting was organized in the City Hall.

Vice President A. C. GOODELL, JR., in the chair.

The records of the preceding meeting were read. Donations were announced to the Museum and Library.

Letters were read from the following:

S. H. Scudder, Sect. Boston Soc. Nat. Hist.; H. A. Bellows, Concord, N. H.; A. W. Dodge, Hamilton; Dr. A. S. Packard, Jr., Boston; C. G. Brewster, Boston; H. A. Purdie, Boston; Samuel Jillson, Feltonville, Mass.; Prof. Joseph Henry, Sect. Smithsonian Institution; Prof. Theo. Gill, Smithsonian Institution; B. Westermann & Co., New York; Ezra Cleaves, Beverly; Mrs. K. N. Doggett, Chicago, Ill.; Hugh Wilson, Salem; A. Lackey, Marblehead; Mrs. P. A. Hanaford, Reading; I. P. Langworthy, Boston; Rev. Geo. D. Wildes, Salem; Paul J. Beckford, Salisbury, relating to the forwarding of specimens and general business: W. M. Hunting, Fairfield, N. Y.; E. S. Morse, Gorham, Me.; Prof. O. P. Hubbard, Dartmouth College; Prof. S. F. Baird, Smithsonian Institution; Dr. Julius Homberger, New York; Prof. E. A. Verrill, Norway, Me.; Capt. Alpheus Hyatt, Gorham, Me.; H. B. Rice, Boston; Surgeon B. G. Wilder, 55th Mass. Vols.; N. S. Shaler, Cambridge; W. Bower, New York; A. C. Goodell, Jr., Salem; W. A. Smith, Worcester, relating to the publications: James P. Kimball, New York, accepting membership.

Joseph P. Cloutman, of Salem, was elected a Resident Member.

Mr. F. W. Putnam, of Salem, was called on for an account of the forenoon ramble. The various specimens that had been collected were displayed on the table, and Mr. Putnam took them up in order. He first showed a fish bone, and explained how from one bone, hair, tooth or scale the character of the living animal could be determined, the analysis in the present case proving the specimen to be part of the jaw of the monk fish (*Lophius*). He next showed several specimens of sandlances (*Ammodytes*), which bury themselves in the sand, when thrown up by the waves, and remain till the next tide allows them to return to their native element. These fishes often lie at the bottom of the water, apparently dead, but on being disturbed revive and become as active as ever. A bottle of minnows was next exhibited and their characteristics explained. The next object in order was the skull of a cat, picked up on the beach, which was interesting from the very extreme age indicated by the teeth, many of which had dropped out, and the cavities become closed. The horse-shoe, or king crab, was next taken up. These were not the animals themselves, but only the shells, the tenants having vacated on their quarters becoming too close for them, a new and larger shell being secreted in a short time. They also cast out the lining of their stomachs. These animals are among the lowest of their class, approaching the fossil trilobite. The sand flea was also referred to as a proper crustacean. The sea urchin was exhibited as a specimen of the radiates, and shown to be in its structure closely allied to the starfish. A black body about two inches long, with prongs projecting from the corners, and which is popularly supposed to be a seaweed bladder, was explained to be the egg case of a skate. The fish attains its perfect form in this case, being supplied with water during its entire growth by means of the four tubes or prongs.

Dr. Henry C. Perkins, of Newburyport, was next called on. He said he came to learn, not to teach, but still would not be selfish. He thought the society had made a collection of all the specimens the waters of this region afforded. He once had a dredge made, and used for several years by a boatman, for the purpose of fishing up, if possible, some new species not found on the shore, but succeeded in finding only one—an arctic shell. He had been interested in watching an excavation in order to study the various strata and other objects of interest. The hill where the observatory stood during the last war, had changed from a northwest to a northeast slope. The sand resembles the Plum Island sand, and at the present time the drought had reached five feet, that being the point where the first indications of moisture were found. When the "James Mills" excavated the hill on the turnpike for a reservoir, they found at a depth of five feet pine logs three feet in diameter at a locality known by tradition as the "Pine Swamp." Lower still, stratified sand was found, and five feet lower, a trunk of a tree within one foot of water. In searching for organic remains he had found gravel cemented to larger stones by lime which had apparently percolated through the strata above from shells. He also referred to the change in the channel in the river, and to the storm which cut off a mile of Salisbury beach, making a channel for the largest ships.

Rev. A. E. P. Perkins, of Ware, made some remarks on geology. He thought geologists were often mistaken in deducing the age of formations, for, owing to causes which we did not understand, the alluvial formation often accumulated in a hundred years as much, as at the slow rates sometimes observed, would indicate ten thousand years. In his native town, not yet a century old, a certain location was known as the beaver dam though no traces of the dam were found or known to the present generation, till, on digging a ditch, it was discovered four feet below the surface, which proved that that depth of alluvium had accumulated in a hundred years.

He then made some remarks on the migrations of birds. He included in this term not only the annual migrations but the permanent change of habitat. Birds often appeared in great numbers in a region to which they had previously been strangers; and, on the other hand, sometimes disappeared entirely from their accustomed haunts. There were many birds in our woods which not one man in ten had ever seen, whose song could be detected by an experienced ear, but never heard by the chance passer by. He instanced the Indigo bird as an example.

Hon. Asahel Huntington, of Salem, gave an interesting reminiscence of Newburyport, and his early acquaintance with many of the prominent divines, physicians and lawyers. He highly eulogized Miss Gould, the poetess, and her father, Capt. Benjamin Gould, who took part in the Revolution and was wounded at the battle of Lexington. He built the house in which Mr. H. was born, and the first rudiments of his education he received in a school taught by a sister of the poetess. The first of Miss Gould's famous series of epitaphs was written for him, at his suggestion, in reply to her assertion that he would kill himself smoking. She complied, and wrote the epitaph off hand, together with some half dozen others the same evening. This was her first attempt at poetry.

Col. Eben F. Stone, of Newburyport, was next called upon. He said that being a new member of the Institute he came to hear, not to talk. His studies had been in other directions than science-ward. He felt the necessity of science—of a knowledge of nature to make his walks more agreeable. He had learned something; and did not believe that the study of science destroyed the poetry and charms of nature.

Vice President Goodell, Chairman of the Committee to report upon the death of Rev. Mr. Barden, presented the following resolutions :

Resolved: That in the recent death of the Rev. Stillman Barden of Rockport, the Institute deplores the loss of a sincere lover of science, and an active and zealous worker in its cause; that it is peculiarly painful to the survivinl members of the Institute to reflect that its meetings wig no longer be enlivened by his presence, nor its memberll encouraged by his ever cheerful voice and his genias manners.

Resolved: That these Resolutions be entered on the records of the Institute, and that the Secretary cause a copy thereof to be sent to the family of the deceased.

The adoption of the resolutions was moved by Dr. Wheatland, and seconded by Rev. Willard Spaulding, of Salem, who spoke with much feeling and earnestness in eulogy of the deceased. The resolutions were unanimously adopted.

The members of the Institute having received and accepted a polite invitation from Hon. Caleb Cushing to visit his house, the meeting adjourned for that purpose, after passing votes of thanks to the City authorities for the use of the Hall, to Hon. Caleb Cushing, and to Rev. Dr. Spalding, and other citizens, for their courtesies and attentions.

Repairing to Mr. Cushing's fine residence, the company were kindly greeted by the host, who not only opened all his rooms for their inspection but also entertained them with a generous hospitality, entirely unexpected, and not often bestowed by any distinguished gentleman upon so numerous a party, principally entire strangers. The privilege of such a reception may be in some measure estimated, when it is stated that Mr. Cushing has one of the finest and most extensive private collections of rare paintings to be found in the United States. They include many celebrated works of the old Spanish masters, and other valuable specimens, not omitting some of the best of Chinese art, obtained by Mr. Cushing during his various sojourns in Mexico, and Europe, and in the Oriental World. The collection comprises more than seventy distinct pieces, of different sizes, and a variety of subjects, many of them of great historical interest and value. He also possesses some choice statuary and several fine family portraits. The examination of these splendid works of art afforded the crowning pleasure of the day.

Mr. Superintendent Prescott furnished an extra train for the return trip, and the party reached home safely, highly delighted with the closing excursion of the season.

Additions to the Museum and Library during July, August, and September, 1865.

TO THE NATURAL HISTORY DEPARTMENT.

BY DONATION.

ALLEN, J. F. Salem. Larvæ and Imago of *Ctenucha grata* from the Grape vine.

BOARDMAN, GEO. A. Milltown, Me. Embryos of the Sheldrake, Ruffed Grouse and Loon, from Milltown. Smelts, *Osmerus sp?*, and *Crangon sp?*, from St. George River. 2 Snakes and a Mineral, from near Milltown.

BOWDOIN, DR. W. L. Salem. Head and feet of a large Turtle, *Chelydra serpentina*, from a Lake in N. H.

CARLEN, SAMUEL Salem. *Syngnathus Peckianus*, from Salem Mill Pond.

CONGDON, MISS EUNICE New Bedford, Mass. Fossil *Astrea*, Vertebra of a *Cetacean*, from York River, about 2 miles from Yorktown, Va. Specimens of Cotton plant, from Virginia. Seed Vessels of a species of *Asclepias*, from Yorktown, Va. 30 specimens, 10 species, *Fossil shells*, from the Bank of James River, Va., near Allen's Landing, about 6 miles from Yorktown.

CLOUTMAN, WM. R. Salem. Specimen of a Beetle, from Salem.

COOKE, C. Salem. Coleoptera, from Pond Lily leaves. 48 specimens, 16 species, Insects, from Reading. Parasites, from the intestines of the Golden-winged Woodpecker.

COOKE, C. and PICKMAN, H. D. Salem. Collection of Insects and Fresh water Fishes, from Rye, N. H.

COVILL, T. N. Salem. 200 specimens of *Pinnotheres ostreum* Say, Oyster Crab.

CREELMAN, MRS. B. C. North Beverly. Specimen of *Solen ensis*, from Salisbury Beach.

CROSS, HENRY J. Salem. *Ascidians*, from the North River flats.

DODGE, A. W. Hamilton. Living Specimen of *Lasiurus noveboracensis* Tomes, Red Bat, from Hamilton.

EDWARDS, CHARLES Salem. Specimen of *Strombus*, from Africa.

EMERTON, JAMES H. Salem. 256 specimens, 60 species, Insects, from North Andover. *Salamander erythronota*, from the Gloucester woods. 36 specimens, 29 species, Insects, from Georgetown.

FELT, S. Q. Salem. *Tropidonotus sauritus*, and a Mineral, from North Andover. Head of an Antelope, from Sierra Leone, W. Africa.

FOOTE, REV. HENRY W. Boston. Collection of Minerals, Shells, Corals and Seeds, from various localities.

HEATH, N. Salem. 105 specimens Insects and Spiders, from Salem.

HINES, MRS. Salem. A Glow-worm (living specimen) *Lampyris noctiluca* (*female larva*), from Boston.

HOOPER, NATHANIEL M. Salem. Nuts, from Cayenne, S. A. Albert Coal, from Hillsboro, N. B.

JOHNSON, DANIEL H. Salem. Clam, *Mya*, with a double shell, from Ipswich.

KIMBALL, JAMES Salem. 6 specimens of "*Spanworm Moths*," from New York.

LANDER, MISS L. Salem. Large Moth, from Salem.

LARABEE, EBEN L. Salem. A. Large collection of Sponges, from under the draw of Beverly Bridge.

LOVETT, EDMONDS Beverly. Skin of a Leopard, from West Africa.

MACK, DR. WM. Salem. Specimen of *Tenia solium*.

MASON,—— Jamestown, N. Y. Silver Ore, from Colorado Territory.

MCILWRAITH, THOMAS Hamilton, C. W. Skins of *Lanius excubitoroides* and *Plectrophanes lapponicus*, from Hamilton, C. W.

NELSON, AUGUSTUS Georgetown. Clay, from Georgetown.

NELSON, HENRY A. Georgetown. Pupa, Imago and Parasite of the *Tent Caterpillar*.

NORWOOD, PROF. J. G. Columbia, Mo. A collection of 687 specimens, 298 species of Western Fossils. Identified.

OSGOOD, MRS. CHAS. Salem. Specimens of a dipterous Insect, from Salem.

PALFREY, CHAS. W. Salem. 5 Eggs of the Mocking Bird, *Mimus polyglottus*.

PICKMAN, H. D. Salem. Dragon-fly, from Salem.

PORTER, E. J. Salem. Specimen of *Epiera riparia*, from Salem.

PRESTON, JOHN Georgetown. Several Minerals, from various localities.

PUTNAM, C. A. Salem. *Dipterous larva*, from the Canal of Naumkeag Mills, Salem. Star-nosed Mole, *Condylura*, from Lawrence. 4 Trout, *Salmo fontinalis*, from the Aqueduct Fountains in South Danvers. 6 *Osmerus*; 2 *Ctenolabrus*; 2 *Morrhua* (young); *Phycis* (young); *Platessa*, from Salem Harbor.

PUTNAM, F. W. Salem. *Storeria Dekayi* and *Tropidonotus sauritus*, from North Andover. *Fundulus multifasciatus*, from Andover Pond. Gall-flies and parasites in several stages, from the Galls on the Wild Rose.

ROBINSON, ASA P. New York. Skin of *Rana Catisbiana*, from Lake Umbagog, Me.

RUSSELL, MISS M. A. Salem. Specimen of Walking stick, *Spectrum Femoratum*.

SAUNDERS, CAPT. O. H. Salem. Lead Ore and Lime, from Island Pond, Canada.

SMITH, SAMUEL H. Salem. Specimen of *Platyphyllum concavum*, Katy-did, from Holmdil, N. J.

SPRINGFIELD CITY LIBRARY MUSEUM, by S. STEBBINS. 500 specimens Spiders, 2 specimens *Polyommatus porsena*, from Springfield.

STEBBINS, S. and BENNETT, C. W. Springfield. Specimen of Rattlesnake, *Crotalus durissus* Linn., from Mt. Tom.

STILES, FREDERIC Topsfield. White Rat from Topsfield.

STONE, ALFRED Providence, R. I. Nest and young of *Vespa maculata*. Nest and young of *Polistes sp?*, from Providence, R. I.

TELLKAMPF, DR. A. New York. *Ascidia nov. sp.*, from Huntington Bay, Long Island.

TENNEY, PROF. S. Poughkeepsie, N. Y. Copperhead Snake, from Mt. Holyoke.

TRUE, JOSEPH Salem. Stone bored by Shells, from the Grand Banks. Insects, from Salem.

UPTON, CAPT. GEORGE Salem. 2 specimens (skins) of Birds, and specimens of Polyzoa, from the Grand Banks?

WEBB, CAPT. BENJ. Salem. Specimen of *Spondylus*, from the West Indies. 2 specimens of Coral and 4 of Minerals, from various localities. Tree Toad, from China?

WHEATLAND, MISS M. G. Salem. Minerals, from the Banks of the Genessee River, Rochester, N. Y.

WILSON, CHARLES S. Salem. Specimen of the Hoary Bat, *Lasiurus cinereus* Allen, from Salem.

WILSON, MRS. THOS. Salem. Nest of the Chimney Swallow, from Salem.

WOODS, HENRI N. Rockport. *Fistularia serrata* Bloch, from Rockport Harbor.

TO THE HISTORICAL DEPARTMENT.

BY DONATION.

ABBOTT, JOHN Beverly. Leaves of the Charter Oak.

ALLANSON, LIEUT. J. S. Marblehead. 4 Confederate Buttons. 2 Confederate Torpedo caps. Several pieces of Confederate fuse. $10, and 50 cent script of Confederate States.

CARPENTER, J. S. Salem. $1, $500 and two $100 (different issues) Confederate paper currency.

CONGDON, MISS EUNICE New Bedford. 3 balls and a fragment of a shell picked up outside the Fort, soon after the Confederate troops evacuated Yorktown, Va. Slate and pieces of Brick from the ruins of William and Mary College, after the battle of Williamsburg, Va. A piece of the Tombstone of "Major William Gooch, dyed Oct. 29, 1650," near the spot where Cornwallis surrendered to Washington. A piece of Window glass, from the house in which Cornwallis had his Head Quarters in the Fort at Yorktown, Va. Lath, Plastering, and Moss from the roof of the "*Capitulation House*," Head Quarters of Gen. Washington, and near which Gen. Cornwallis surrendered, Oct. 19, 1784, about 1½ miles from the Fort at Yorktown, Va.

DENSLOW, W. W. New York. Revolutionary button of the 57th Regiment, British Army, found on Washington Heights.

FOOTE, CALEB Salem. $10 note Confederate currency.

FOOTE, REV. HENRY W. Boston. 13 Plaster Medallions. Seeds and leaves, from various Historical places, and other specimens.

GOLDTHWAITE, JOSEPH A. Salem. A collection of North Carolina paper currency.

GRANT, LIEUT. FRANKLIN Salem. Cutlass taken from the wreck of the Confederate Steamer Merrimac, at Newport News. Piece of one of three Muskets stacked over the magazine of Ft. Fisher at the time of the explosion.

HOTCHKISS, HENRY L. New Haven, Ct. Photographic views of Ike Marvel's House; Temple Street, New Haven; Library Building and Alumni Hall, Yale College; Hillhouse Avenue, New Haven; Prof. B. Silliman, Sen., and President Woolsey.

LONG, ISAAC M. Salem. 1628 "Patriotic Envelopes" collected during the first part of the Rebellion.

LOVETT, EDMONDS Beverly. 3 Native swords, from the West Coast of Africa.

ORDWAY, COL. ALBERT Richmond, Va. 120 specimens, different denominations and issues, Confederate paper currency.

PITMAN, AUGUSTUS P. Salem. Palmetto Flag.

ROBERTS, DAVID Salem. Confederate paper currency.

SHORT, JOSEPH Salem. Various relics from the Battle Field of Gettysburg.

TENNEY, GORHAM D. Georgetown. Indian Arrow Head, from Georgetown.

WEBB, CAPT. BENJAMIN Salem. Chinese toy.

WILLIAMS, W. A. Salem. Indian relics, consisting of a stone pot, stone chisel, stone arrowheads and a twisting-stone, also a few small bones of a skeleton, and a piece of Red Ochre, taken from an Indian grave on Salem Neck, under the embankment of Ft. Pickering.

TO THE LIBRARY.

By Donation.

Andrews, Mrs. James H. Endicott's Memoirs of John Endicott, 1 vol., 4to, Salem, 1847.

Atwood, E. S. Atwood's Discourse on Lincoln, 8vo, pamph., Salem, 1865.

Barrande, Joachim A. Paris. (Through Smithsonian Institution.) Defense des Colonies par J. Barrande, 1 vol., 8vo, Paris, 1865.

Batchelder, Mrs. John H. The Last will and Testament of Capt. Miles Standish, Broad-sheet.

Board of Agriculture of Lower Canada. Prize List for the Exhibition at Montreal, Sept., 1865, 8vo, pamph.

Brooks, Henry M. Fidler's observations in United States and Canada, 1 vol., 12mo, New York, 1833.

Brunet, Le Abbé Ovide Quebec. Catalogue des Plantes Canadiensis by Brunet, 1st Liv. 8vo, pamph., Quebec, 1865.

Chapman, John Attwood's Discourse on Lincoln, 8vo,, pamph., Salem, 1865.

Chase, Mrs. E. E. U. S. Sanitary Commission, Nos. 87 and 90. The Sanitary Reporter for May 15, 1865. The Sanitary Commission Bulletin for June 1, 1865, and No. 39. U. S. Sanitary Commission pamphlets, 5.

Chase, George C. Friend's Review, 9 Nos. Proceedings of the Alumni Association of Friends Yearly Meeting School, 1865, 8vo, pamphlet.

Chase, George H. Intellectual Symbolism, a basis for Science by Pliny E. Chase, 1 vol., 4to, Phil., 1863. Sanscrit and English Analogues by Pliny E. Chase, 1 vol., 8vo, London, 1860.

Chase, Thomas Haverford College, Penn. The Manuscripts of the Satyricon of Petronius Arbiter, Des. and col. by Chas. Beck, 1 vol., 4to, Cambridge, 1863.

Congdon, Eunice New Bedford. De Obligatione Conscientiae Praelect, Decem a Roberto Sandersono, 1 vol., 12mo, Londini, 1719.

Couper, William Quebec, C. E. Fraser's Journal relating to the Seige of Quebec in 1759, 8vo, pamph.

Curwen, George R. Church Review, vol. xv, Nos. 1, 2, 3, 4. Miscellaneous pamphlets, 16.

Denslow, W. W. New York. Plays, by Barry Cornwall, H. H. Milman, James Haynes and David P. Brown, 1 vol., 8vo.

Drowne, Charles Troy, N. Y. Annual Register of the Rennselaer Polytechnic Institute, 1865, 8vo, pamph.

Eliot, John F. Boston. Reports of Mass. Humane Society, 5.

FABENS, JOSEPH WARREN The Uses of the Camel, a paper by J. W. Fabens, 8vo, pamphlet, New York, 1865.

GILL, THEODORE Washington. Descriptions of New Species &c., by Theo. Gill, in 17 pamphlets.

GREEN, SAMUEL A. Boston. Warder and Catlett's Account of the Battle at Young's Branch or Manassas Plain, July 21, 1861, 16mo, 1 vol., Richmond, 1862. New Testament, 1 vol., 16mo, Atlanta, Ga., 1862. 25 Pamphlets printed in Richmond relating to the Confederacy. Report of the School Committee of Boston, 1864, 1 vol., 8vo, Boston, 1865. The Boston Business Directory, 1863—4, 1 vol., 12mo. 49 Miscellaneous pamphlets.

HOLMES, JOHN C. Detroit, Mich. Michigan School Report and Laws, 1863, 1 vol., 8vo. Michigan School Report, 1864, 1 vol., 8vo. 3d An. Rep. of Michigan Board of Agriculture, 1 vol., 8vo, Lansing, 1864. Elford's Marine Telegraph, 1 vol., 8vo, Charleston, 1823. Warren's Ten thousand a year, 1 vol., 8vo, Phil., 1841. Mitchell's United States, 1 vol., 8vo, Phil., 1834.

HOTCHKISS, SUSAN V. New Haven, Conn. Ten pamphlets and College Exercises relating to Yale College.

HUGUET-LATOUR, L. A. Montreal. Journal of Education, vol. VII, Nos. 7, 8, 9, 10, 11, 12, 4to, Montreal, 1863. Journal de Institution Publique, vol. VII, Nos. 9, 10, 11, 12, 4to, Montreal, 1863.

LANDER, W. W. A collection of blanks used in the department of the Commissary of Subsistence, U. S. Army.

LANGWORTHY, ISAAC P. Philadelphia. 10 Reports of the American Sunday School Union, 8vo, pamph., Philadelphia.

LEE, JOHN C. Stewart's Geography for Beginners (Palmetto Series) 1 vol., 12mo, Richmond, 1864.

LESQUEREUX, LEO Columbus, Ohio. Botanical and Palæontological Report of the Geol. Survey of Arkansas, 8vo, pamph. Palæontological Rep. of Geol. Survey of Kentucky, 8vo, pamph. Musci Boreal-Americani, by Sullivant and Lesquereux, 8vo, pamph. Lesquereux on Coal formations of North America, 8vo, pamph. Lesquereux on Californian Mosses, 4to, pamph. 4th Report of the Geol. Survey of Kentucky, 8vo, 1 vol., Frankfort, 1861. Lesquereux on the Origin of Prairies, 8vo, pamph.

LORD, N. J. Boston Post for April, May and June, 1865.

MACK, SAMUEL E. St. Louis Mo. Edward's St. Louis Directory, 1865, 1 vol., 8vo.

MANNING, ROBERT Perry's Eulogy on Stanley, 8vo, pamph., Salem, 1865.

MEEHAN, THOMAS Philadelphia. The Gardeners' Monthly, vols. 2, 3, 4, 5, 6 and 7, 8vo, Phil., 1860, &c.

Norwood, J. G. Columbia, Mo. 1st and 2d Annual Reports of Geol. Survey of Missouri, by G. C. Swallow, 1 vol., 8vo, Jefferson City, 1855. Catalogue of Univ. of Missouri for 1862, 3, 4 and 5, 8vo, pamph.

Osgood, George P. Autocrasy in Poland and Russia by Julian Allen, 1 vol., 12mo, New York, 1854.

Owen, Richard New Harmony, Ind. Report on the Mines of New Mexico, by Owen and Cox, 8vo, pamph.

Paine, Nathaniel Worcester. Bullock's address at Worcester, June 1, 1865, on A. Lincoln, 8vo, pamph.

Palfray, Charles W. An. Rep. of Adj. Gen. of Missouri, 1863, 1 vol., 8vo.

Perkins, Henry Philadelphia. The Soldier's Guide in Philadelphia, 1865, pamph.

Phillips, Stephen H. Sectional Maps of Farming and Wood Land for sale by Illinois Central R. Road, 8vo, pamph., Chicago, 1865. Lamark, Historie Naturelle des Animaux Sans Vertebres, vol. 1, pt. 1, and vol. 2, 8vo, Bruxelles, 1837. Walter S. Newhall, a memoir, 1 vol., 8vo, Phil., 1864. Dietrichsen and Hannay's Royal Almanack for 1861, 8vo, pamph., London, 1861. Ocean Telegraphing by S. F. Van Choate, 8vo, pamph., Cambridge, 1865.

Pitman, Augustus P. Brevard's Digest of Pub. Statute Laws of S. Carol., vol. 2, 8vo, Charleston, 1814. Acts of Assembly of S. Carol., 1801 to 1804, 8vo, 1 vol., Charleston. Several unbound Manuscripts.

Pulsifer, David Boston. The State House in Boston, Mass. by David Pulsifer, 12mo, pamph., Boston, 1865.

Randall, Stephen Providence, R. I. Aspinwall's remarks on the Narragansetts patent, 2d ed., 8vo, pamp., Providence, 1865.

Ropes, Charles A. Trow's New York City Directory, vol. 75, 1 vol., 8vo, New York, 1862.

Safford, Joshua Spiritual Songs, 1 vol., 12mo, Boston, 1787.

Stearns, George L. Boston. 26 Pamphlets.

Stevens, Miss Caroline Rochester, N. Y. Rochester Directory for 1847—8, 1 vol., 12mo. Several Newspapers.

Trübner & Co., London. Trübner's Amer. and Orient. Literary Record, vol. 1, Nos. 2 and 3.

Waters, H. F. G. Regulations of Med. Dep't of Confed. States Army, 8vo, pamph., Richmond, 1861.

Waters, J. Linton Chicago, Ill. The Prairie Chicken, 1864, 1 vol., 4to, Tilton, 1864—5. Monthly, under the auspices of the late Mrs. Kirkland.

Webb, Benjamin Howison's Dictionary of the Malay Tongue, 1 vol., 4to, London, 1801.

Wheatland, Martha G. Daily Evening Transcript, July, 1864, to July, 1865, 2 vols., folio.

WHEATLAND, STEPHEN G. Roll of Students of Harv. Coll. in the Army and Navy during the Rebellion, 12mo, pamph., 1865. Porcellian Catalogue, 1865, 8vo, pamph.

WIGGIN, J. K. Boston. A. L. Stone's, Discourse on A. Lincoln, April 16, 1865, 8vo, pamph.

WILLIAMS, HENRY L. 12 Rail Road Reports.

BY EXCHANGE.

AMERICAN ANTIQUARIAN SOCIETY. Proceedings of Meeting April 26, 1865, 8vo, pamph.

AMERICAN GEOGRAPHICAL AND STATISTICAL SOCIETY. Proceedings, pages 117—176, 8vo, pamph, New York, 1865.

AMERICAN, PHILOSOPHICAL SOCIETY. Proceedings, No. 73, pamph., Phil., 1865.

BOSTON PUBLIC LIBRARY. A Memorial of Joshua Bates, from the city of Boston, 1 vol., 8vo, Boston, 1865.

CALIFORNIA ACADEMY OF NATURAL SCIENCES. Proceedings, vol. 3, pt. 2, 1864, San Francisco, 1864.

CANADIAN INSTITUTE. The Canadian Journal for July, 1865.

CHICAGO HISTORICAL SOCIETY. Brown's Historical Sketch of the early movements in Illinois for legalizing Slavery, 8vo, pamph., Chicago, 1865.

DARTMOUTH COLLEGE LIBRARY. Catalogus Collegii Dartmuthensis, 1864, 8vo, pamph. Catalogue of Dartmouth College for 1864—5, 8vo, pamph.

EDITORS. The Gardener's Monthly for July, Aug. and Sept., 1865.
American Journal of Science for July and Sept., 1865.
Savannah Daily Herald.
Florida Union.
Historical Magazine for July and Aug.
American Journal of Ophthalmology, vol., 2, No. 2.
The Home Evangelist.
North American Review, July, 1865.
The Essex Banner.
Haverhill Gazette.
Lawrence American.
Salem Observer.
South Danvers Wizard.
Lynn Weekly Reporter.

IOWA STATE HISTORICAL SOCIETY. The annals of Iowa for July, 1865, 8vo, pamph.

LONG ISLAND HISTORICAL SOCIETY. 2d Annual Report, 8vo, pamph., Brooklyn, 1865.

MUSEUM OF COMPARATIVE ZOÖLOGY AT CAMBRIDGE. Illustrated Catalogue of, No. 1, 8vo, pamph., Cambridge, 1865.

NEW ENGLAND HISTORIC—GENEALOGICAL SOCIETY. N. E. Hist. Gen. Register for July, 1865. Eulogy on A. Lincoln before N. E. Hist. Gen. Soc., May 3, 1865, 8vo, pamph.

NEW YORK CHAMBER OF COMMERCE. An. Rep., 1864—5, 1 vol., 8vo, New York, 1865.

NEW YORK LYCEUM OF NATURAL HISTORY. Annals vol. VII, Nos. 1, 9, vol. VIII, Nos. 2, 3, 4 and 5.

NEW YORK MERCANTILE LIBRARY ASSOCIATION. 44th Annual Report, 8vo, pamph.

PHILADELPHIA ACADEMY OF NATURAL SCIENCES. Proceedings No. 2, for April, May and June, 1865.

QUEBEC LITERARY AND HISTORICAL SOCIETY. Transactions, Session of 1864—5, 8vo, pamph., Quebec, 1865.

REDWOOD LIBRARY AND ATHENÆUM. Report Made Sept. 28, 1864, 8vo, pamph., New York, 1864.

YALE COLLEGE LIBRARY. Catalogus Collegii Yalensis, 1865, 8vo, pamph. Obituary Record of the Graduates of Yale College, July 26, 1865, 8vo, pamph.

TUESDAY, OCTOBER 3. Adjourned Regular meeting.

Vice President ALLEN in the chair.

Samuel Q. Felt, of Salem, was elected a Resident Member.

MONDAY, OCTOBER 16. Regular meeting.

N. WESTON, JR., in the chair.

The Secretary read by title the following communication:—"*Prodrome of a Monograph of the Pinnipedes.*" By Prof. Theodore Gill, of the Smithsonian Institution.

James N. Estes, of South Danvers, was elected a Resident Member.

MONDAY, NOVEMBER 6. Regular meeting.

Dr. GEORGE B. LORING in the chair.

Letters were read from:—

T. McIlwraith, Hamilton, C. W.; M. S. Bebb, Washington, D. C.; W. H. Niles, Cambridge; W. E. Endicott, Canton; H. C. Perkins, Newburyport; N. S. Shaler, Cambridge; Stephen D. Poole, Lynn; Prof. A. E. Verrill, New Haven, Ct.; W. M. Hunting, Fairfield, N. Y.; L. L. Thaxter, Boston; B. P. Mann, Concord; E. E. Barden, Rockport; Prof. C. E. Hamlin, Waterville, Me.; R. E. C. Stearns, San Francisco, Cal.; Prof. J. Wyman, Cambridge; Prof. W. B. Rogers, Boston; Julius Silversmith, New York, N. Y.; A. B. Kendig, Marshalltown, Iowa; W. O. White, Keene, N. H.; W. O. Currier, Providence, R. I.; E. W. Hervey, New Bedford; W. W. Denslow, New York, N. Y.; John Bolton, Portsmouth, Ohio; American Philosophical Society; Joseph Blake, Gilmantown, N. H., relating to the publications: E. Suffert, Matanzas, Cuba; Dr. J. C. Puls, Ghant, Belgium; Prof. R. Owen, Bloomington, Ind.; James Lewis, Mohawk, N. Y.; Smithsonian Institution; G. L. Stearns, Boston; W. Wallis, Salem; Dr. S. A. Green, Boston; A. W. May, Boston; Mrs. E. E. Chase, Salem; J. Linton Waters, Chicago, Ill.; Rev. C. F. Barnard, Boston, relating to the transmission of specimens and books: B. O. Peirce, Beverly; Prof. James Hubbert, Richmond, C. E.; Prof. A. E. Verrill, New Haven, Ct.; Geo. C. Huntington, Kelley's Island, Ohio; E. T. Cresson, Philadelphia, Pa.; Horace Mann, Concord; A. E. Kursheedt, Cincinnati, Ohio; Mrs. P. A. Hanaford, Reading; J. F. Allen, Salem; W. B. Trask, Boston; John Krider, Philadelphia, Pa.; Dr. Wm. Stimpson, Chicago, Ill.; W. J. Howard, New York, N. Y.; Henri N. Woods, Rockport; Chas. A. Emery, Springfield; G. A. Boardman, Milltown, Me.; Hon. C. Cushing, Newburyport; I. L. Gosling, New York, N. Y.; H. A. Bellows, Concord, N. H.; James W. Perkins, Salem; J. C. Holmes, Detroit, Mich.; Prof. J. G. Norwood, Columbia, Mo.; Wm. Merritt, Salem; Capt. D. H. Johnson, Jr., Salem; Dr. H. C. Perkins, Newburyport; Prof. S. Tenney, Poughkeepsie, N. Y.; T. A. Cheney, Havana, N. Y.; E. E. Barden, Rockport; Hon. O. P. Lord, Salem; F. W. Putnam, Salem; Hon. A. Huntington, Salem, on general business: New York Lyceum of Natural History; Buffalo Society of Natural Sciences; New York Historical Society; Trustees of Dartmouth College; New England Historic-Genealogical Society; Museum Comp. Zoölogy; Albany Institute; Maine Historical Society, acknowledging the receipt of publications: Prof. J. G. Norwood, Columbia, Mo.; Josiah Stickney, Boston; A. T. Mosman, Beverly, accepting membership.

Donations to the Library and Museum were announced.

Adjourned to Tuesday evening, Nov. 14.

WEDNESDAY, NOVEMBER 8. Stated meeting.

N. WESTON, JR., in the chair.

Adjourned to Tuesday evening, Nov. 14.

COMMUNICATIONS, VERBAL, BY.

Communications, Written, by.

Members Elected, Resident.

MEMBERS ELECTED, CORRESPONDING.

PUBLICATIONS OF THE ESSEX INSTITUTE.

For Sale by the Secretary, or exchanged for the Publications of other Societies.

JOURNAL of the Essex County Natural History Society, 8vo. 1836 . . . $0 50

PROCEEDINGS of the Essex Institute, 8vo. Vol. I. 1848—1856 2 00

" " " " " " II. 1856—1858 . . . 2 00

" " " " " " III. 1858—1863 . . . 2 00

" " " " " " IV. 1864 . . (present volume)

HISTORICAL COLLECTIONS of the Essex Institute, 8vo. Vol. I. 1859 . . 2 00

" " " " " " " II. 1860 . . 2 00

" " " " " " " III. 1861 . . 2 00

" " " " " " " IV. 1862 . . 2 00

" " " " " " " V. 1863 . . 2 00

" " " " " " " VI. 1864 (present vol.) 2 00

COLE'S Infusoria of Salem. Pamphlet 8vo. 1853 . . . . 0 15

WHITE'S Covenant of the First Church. Pamphlet 8vo. 1856 . . . . . 0 10

STREETER'S Account of the Newspapers and other Periodicals published in Salem. Pamph. 8vo. 1856 . . . . 0 15

ENDICOTT'S Account of Leslie's Retreat, 8vo. 1856 . . . . 0 25

FOWLER'S Account of the Life, Character, &c., of the Rev. Samuel Parris, and of his connection with the Witchcraft Delusion of 1692. Pamph. 8vo. 1857 . . 0 15

WHITE'S Memoir of the Plummer Family. Pamph. 8vo. 1857. 0 15

DEDICATION of Plummer Hall. Pamphlet 8vo. 1857 . . . . . 0 30

WEINLAND'S Egg Tooth of Snakes and Lizards. Pamph. 8vo. with a plate. 1857 0 15

ENDICOTT'S Account of the Piracy of the ship Friendship of Salem in 1831. Pamph. 8vo. 1859 . . . . 0 15

THE WEAL-REAF, a Record of the Essex Institute Fair. Pamph. 8vo. 1860 . 0 30

WHITE'S New-England Congregationalism, 8vo. 1861 . . . . 1 00

UPHAM'S Memoir of Gen. John Glover of Marblehead. Pamphlet 8vo. 1863 . 0 50

BRIGGS' Memoir of D. A. White. Pamphlet, 8vo, 1864. . . . . . . 0 30

NATURALISTS' DIRECTORY.

The first part of the Naturalists' Directory, containing the names of the American Naturalists, will be issued with the next number of the Proceedings.

For information relating to the Directory, address

F. W. PUTNAM,
Essex Institute,
Salem, Mass.

State that have not been noticed at Springfield. The list will be seen to consist mainly of those Water Birds that frequent the coast and are not found far inland, with a few rare or accidental visitors. No species is admitted of which there is not good evidence of its capture in the State; and when the species is extremely rare, the authority is cited on which it is inserted. Consequently some species that have been attributed to Massachusetts, from their occurrence in adjoining States, though probably to be found here as rare visitors, and are thus mentioned, are not counted as a part of the list; very careful observers will, doubtless, yet detect most of them here.

1. *Cathartes atratus* Less. Black Vulture. Accidental. One was obtained in Swampscott, in November, 1850. (S. Jillson, Proc. Ess. Inst., Vol. I, p. 223 —Brewer's N. Am. Oölogy, pt. I, p. 5.) Another was taken the past season, Sept. 28, at Gloucester, by Mr. William Huntsford. (A. E. Verrill.)

2. *Cathartes aura* Ill. Turkey Vulture. Accidental. Two were taken in the State in 1863. (E. A. Samuels, Agr. Mass., 1863, Secy's Rep , App., p. XVIII.)

3. *Falco candicans* Gm. Jer Falcon. Accidental in winter. One was shot at Sekonk Plains, about 1840. (S. Jillson, Proc. Ess. Inst., vol. I, p. 226.) Has been seen here by Nuttall and others.

4. *Aquila canadensis* Cass. Golden Eagle. Extremely rare; but few recorded instances of its capture in the State. (Lynn, S. Jillson, Proc. Ess. Inst., vol. I, p. 203. Lexington, Dr. Kneeland, Proc. B. S. N. H., vol. V, p. 272. Near Boston, Brewer, N. Am. Oöl,. pt. I, p. 45.—Upton, Agr. Mass., 1859, Secy's Rep., p. 141.)

5. *Syrnium cinereum* Gmelin. Great Cinerous Owl. Occasional in winter. (Marblehead, February, 1831, and January, 1835; S. Jillson, Proc. Ess. Inst., vol. I, p. 204.) Seven were taken in the State during the year ending February, 1843. (Dr. S. L. Abbot, Proc. B. S. N. H., vol. I, pp. 57 and 99.) Two specimens in the Mus. Comp. Zoölogy were obtained in 1848, in the Boston markets, and were *probably* killed in the State.

The Hawk Owl (*Surnia ulula* Bon.) is said by Prof. Emmons to have been seen in autumn. Though I have found no notice of its capture, it is not improbable that it may occasionally occur along the Green Mountains in the Western part of the State.

The Banded Three-toed Woodpecker (*Picoides hirsutus* Gray) has been repeatedly attributed to the State, and may occur as a very rare or accidental winter visitor.

6. *Hylotomus pileatus* Baird. Pileated Woodpecker. "Log Cock." Rare. Driven from most parts of the State by the absence of extensive forests, but is still found in the wooded, mountainous parts of Berkshire County.

The Varied Thrush (*Ixoreus nœvius* Bon.) is said by Prof. Baird, in the Reports on the Pacific Railroad Explorations and Surveys, vol. IX, pp. XXI and 219, to be accidental near Boston, quoting Dr. Cabot (Proc. Bost. So. N. H., vol. III, p. 17) as authority. Dr. Cabot states that a specimen of this species was obtained in *Boston market*, but adds that it was *shot in New Jersey*. This is the only notice I can find respecting this species being found in Massachusetts, either by Dr. Cabot or others.

7. *Oporornis agilis* Baird. Connecticut Warbler. Very rare. Was taken in Berlin, in the summer of 1845. (Dr. S. Cabot Jr., Proc. Bost. So. N. H., Vol. II, p. 63.)

8. *Helmitherus vermivorus* Bon. Worm-eating Warbler. Very rare. Its nest has been found in Cambridge. (Peabody's Rep. Orn. of Mass., p. 312.*)

9. *Helmitherus Swainsonii* Bon. Swainson's Warbler. Audubon states, on the authority of Dr. T. M. Brewer, that one was taken in Massachusetts by Mr. S. Cabot Jr. (Aud. Orn. Biog., vol. V, p. 462.) Mr. Peabody probably alludes to the same specimen (Rep. on Orn. of Mass., p. 213.) Very rare in this State.

10. *Helminthophaga pinus* Baird. Blue-winged Yellow Warbler. Summer visitant. Very rare. (S. Cabot Jr., Proc. B. S. N. H., vol. VI, p. 386.)

11. *Helminthophaga chrysoptera* Baird. Golden-winged Warbler. Summer visitant. Very rare. (S. Cabot Jr., Proc. B. S. N. H., vol. VI, p. 386.) Have seen specimens in the Mus. Comp. Zoöl., Cambridge, that were taken in the State.

---

* Fishes, Reptiles and Birds of Massachusetts.

The Blue Warbler *(Dendroica cœrulea* Baird.*)* is said to be a rare summer visitant, (F. W. Putnam, Proc. Ess. Inst., vol. I, p. 207,) but I have failed to find an authentic instance of its capture in this State. Audubon says it has been taken at Pictou, Nova Scotia, and so may very naturally be expected to occur in Massachusetts.

12. *Wilsonia minuta* Bon. *(Myioioctes minutus* Baird.) Small-headed Flycatcher. This little known and rather doubtful species is said to occur in this State. (Ipswich, Dr. T. M. Brewer; Berkshire County, Prof. E. Emmons. Peab. Rep. Orn. Mass., p. 297.—Salem, T. Nuttall, Man. Orn., vol. I, p. 297.)

The Hooded Flycatcher *(Wilsonia mitrata* Bon.; *Myiodiœtes mitratus* Aud.) may be looked for in this State, as it has been found in Connecticut and New York. Mr. E. A. Samuels, in his recent list of the Birds of Massachusetts, (Agr Mass., 1863, Secy's Rep., App., p. XXII,) gives it as a rare summer visitor.

13. *Pyranga æstiva* Vieill. Accidental. "Two were taken in Lynn, after a severe storm, April 21st, 1852." (S. Jillson, Proc. Ess. Inst., vol. I, p. 224.)

14. *Vireo noveboracensis* Bon. White-eyed Vireo. Summer visitant. Not very uncommon in the eastern part of the State, where it breeds.

15. *Cistothorus (Telmatodytes) palustris* Cabanis. Marsh Wren. Summer visitant. Rare.

16. *Cistothorus stellaris* Cab. Short-billed Marsh Wren. Summer visitant. Not uncommon.

The Blue Gray Gnatcatcher *(Polioptila cœrulea* Sclat.*)* is said by Peabody to be found in Massachusetts, on the authority of Dr. Brewer, (Rep., p. 297.) Having been found in adjoining States,—in New York north of the latitude of Boston, as well as in Nova Scotia, and in Connecticut,—it may be looked for as a rare straggler from its usual habitat. I have been unable as yet to learn of its actual capture in this State.

The Crested Chickadee *(Lophophanes bicolor* Bonap.*)* though mostly a southern species, Audubon states (Orn. Biog., vol. V, p. 472) is common in Nova Scotia, and hence may be expected to occur here.

17. *Parus hudsonicus* Forster. Hudsonian Titmouse.

Occasional or accidental in winter. (Brookline, S. Elliot Green, Peabody's Rep., p. 402.) Resident at Calais, Maine, but not common. (G. A. Boardman, Proc. B. S. N. H., vol. IX, p. 126.)

18. *Centrophanes lapponicus* Kaup. Lapland Longspur. Winter visitant. Occasional, or accidental. (F. W. Putnam, Proc. Ess. Inst., vol. I, p. 210.)

19. *Ammodromus maritimus* Swain. Sea-side Finch. Summer visitant. Common in the salt marshes along the coast, where it breeds.

20. *Ammodromus caudacutus* Swain. Sharp-tailed Finch. Common summer visitant in salt marshes, where it breeds. Have taken it in the marshes of Charles River the last week in October.

21 *Chondestes grammaca* Bon. Lark Finch. Accidental. "One found in Gloucester, about 1845." (S Jillson, Proc. Essex Inst., vol. I, p. 224.)

22. *Euspiza americana* Bon. Black-throated Bunting. Probably rare or occasional. Said to be found here by Nuttall (Man. Orn., vol. I, p. 461). According to Peabody, "is found in high meadows near salt water marshes, from the middle of May till the last of August." (Rep. Orn. of Mass., p. 319.) Mr. E. A. Samuels informs me that he has seen two specimens killed in this State; one was sent him from Woburn. Nuttall, in his account of the notes and habits of this species, as observed here, has described the peculiar song and habits of the Yellow-winged Sparrow (*Coturniculus passerinus* Bon.) with remarkable aptness, which species he evidently mistook for the Black-throated Bunting. Nuttall seems not to have known the Yellow-winged Sparrow, under its proper name, at the time he wrote, and it is difficult to tell what he had in mind when describing its habits and distribution in the breeding season; his description of its song, which he strangely likens to that of the Purple Finch, and of its eggs, being not at all applicable to the Yellow-winged Sparrow. As Nuttall has been the authority chiefly depended on for the occurrence of *Euspiza americana* in this State, I strongly doubted its having been taken here, till assured of the fact by Mr. Samuels.

The Blue Grosbeak (*Guiraca cœrulea* Swain.) may be

looked for as an occasional visitor. Has been found at Calais, Maine, where it is "very uncertain, but common in the spring of 1861." (G. A. Boardman, Proc. B. S. N. H., vol. IX, p. 127.)

23. *Cardinalis virginianus* Bon. Cardinal. Red Bird. Accidental summer visitant, according to Nuttall. (Man. Orn., vol. I, p. 519.) Seen "only at irregular intervals, in the villages on the Connecticut river." (Peabody, Rep. Orn. Mass., p. 329.)

24. *Quiscalus major* Vieill. Boat-tailed Grakle. Accidental. Have heard of one that was killed in Cambridge a few years since. Mr. E A. Samuels tells me that a pair bred in Cambridge in 1861.

25. *Corvus ossifragus* Wils. Fish Crow. An occasional visitor along the southern coast of the State.

26. *Tetrao canadensis* Linn. Spruce Partridge. Accidental. Found in the hemlock woods of Gloucester, in September, 1851. (S. Jillson, Proc. Ess. Inst., vol. I, p. 224.)

27. *Cupidonia cupido* Baird. Pinnated Grouse. Prairie Hen. Nearly extinct in Massachusetts. A few are occasional visitors in the southeastern part of the State, from Long Island, where they still remain. (S. Cabot Jr., Proc. B. S. N. H., vol. V, p. 154.) About thirty years since were quite common in Martha's Vineyard (Audubon Birds Amer., vol. V, p. 101.)

The Wild Turkey (*Meleagris gallopavo* Linn.) is now probably extinct in this State. Within a few years it has been said to occur wild on Mts. Tom and Holyoke; but I can find no authentic instance of its recent capture in this State. The accounts of those recently taken seem to rest on the authority of hunters, who might readily mistake a stray domestic turkey for a wild one, and not on the authority of reliable naturalists. It is well known that the domestic turkey will sometimes take to the woods, assuming the habits of the wild bird; hence these reports may well be received with considerable caution. In winter the wild birds are found in Boston markets, but are brought from distant parts of the country, chiefly from the West.

28. *Garzetta candidissima* Bonap. Snowy Heron. Ac-

cidental. Stragglers have been taken in a few instances. Have seen one that was killed near Boston, in 1862.

29. *Florida cœrulea* Baird. Blue Heron. Stragglers only taken here. There is a specimen in the State Agricultural Cabinet, taken in the eastern part of the State.

30. *Ibis Ordii* Bonap. Glossy Ibis. Occasional; apparently accidental. Have been taken here at irregular intervals. In June, 1830, three were obtained in the eastern part of the State. (Nuttall, Man. Orn., vol. II, p. 88.) Others have been taken. (Cabot, Proc. B. S. N. H., vol. III, pp. 313, 333, 355; vol. IV, p. 346.)

31. *Octhodromus Wilsonius* Reich. (*Ægialitis Wilsonius* Cass.) Wilson's Plover. Occasional in summer. Inserted on the authority of Dr. Brewer, who found them, according to Peabody (Rep. Orn. Mass. p. 360), "abundant at Nahant, in August," 1838.

32. *Ægialeus melodus.* (*Ægialitis melodus* Cab.) Piping Plover. Common visitant, mostly along the seacoast in summer, some breeding.

33. *Hœmatopus palliatus* Temn. Oyster Catcher. Very rare. Has been found in the State by Dr. Brewer. (Peab. Rep. Orn. Mass., p. 358.)

34. *Strepsilas interpres* Ill. Turnstone. Common spring and autumn visitant, along the coast.

The American Avoset (*Recurvirostra americana* Gmel.) and the Black-necked Stilt (*Himantopus nigricollis* Vieill.), from their general distribution, may be looked for in Massachusetts as very rare species.

35. *Phalaropus Wilsonii* Sab. Wilson's Phalarope. Very rare. Found in the State by Audubon. (Birds Am., vol. V, p. 301.)

36. *Phalaropus hyperboreus* Temn. Northern Phalarope. Along the coast; not common.

37. *Phalaropus fulicarius* Bon. Red Phalarope. Occasional visitor, chiefly along the coast, in spring and autumn.

38. *Macrorhamphus griseus* Leach. Red-breasted Snipe. Not very common. Spring and autumn visitant, near the coast.

39. *Tringa canutus* Linn. Ash-colored Sandpiper.

Knot. "Gray Back." Spring and fall; sometimes very abundant in autumn, arriving in August.

40. *Arquatella maritima* Baird. (*Tringa maritima* Brünn.) Purple Sandpiper. "Rock Snipe." On the coast in autumn; not generally common.

41. *Ancylochilus subarquata* Kaup. Curlew Sandpiper. Coast; not common.

42. *Actodromas Bonapartii* Cass. (*Tringa Bonapartii* Schl.) Bonaparte's Sandpiper. Coast in spring and fall; sometimes abundant.

43. *Ereunetes pusilla* Cass. Semipalmated Sandpiper. Common along the coast in spring and autumn.

44. *Limosa fedoa* Ord. Marbled Godwit. Rare passenger in spring and fall.

45. *Limosa hudsonica* Swain. Hudsonian Godwit. Spring and fall. Not common.

46. *Numenius longirostris* Wilson. Long-billed Curlew. Spring and fall. Not common.

47. *Numenius hudsonius* Lath. Hudsonian Curlew Rare. Spring, fall and winter.

48. *Numenius borealis* Lath. Esquimaux Curlew. Spring and fall; occasionally in winter on the coast. Rare.

49. *Rallus crepitans* Gm. Clapper Rail. Rare or accidental. (S. Cabot Jr., Bost. So. N. H., vol. III, p. 326.)

50. *Porzana noveboracensis* —? Yellow Rail. Found in spring and fall; perhaps breeds. Not common.

51. *Gallinula galeata* Bonap. Florida Gallinule. Accidental. Has been taken at Fresh Pond, Cambridge, by Mr. Cabot. (Peab. Rep. Orn. Mass., p. 258.)

52. *Gallinula martinica* Lath. Purple Gallinule. Like the preceding, occurs as a very rare, chance visitor from the south, but is oftener met with. Has been taken but a few times in this State.

53. *Anser hyperboreus* Pallas. Snow Goose. Winter visitant. Not common.

54. *Anser Gambellii* Hartl. White-fronted Goose. Have seen specimens obtained in Boston market that were probably taken in the State.

55. *Bernicla Hutchinsii* Bonap. Hutchin's Goose.

This species is introduced as a bird of Massachusetts with considerable doubt. For its occurrence here, we have the authority of Nuttall, (Man. Orn., vol. II, p. 362) who mentions it as a straggler on our coast,—and of Giraud, who says it is quite abundant some seasons on the coast of Massachusetts. Lindsley, in his Catalogue of the Birds of Connecticut, (Am. Jour. Sc. and Arts, vol. XLIV, p. 249) says it is not unfrequently taken in Connecticut in spring.

56. *Bernicla leucopsis* —? (*Anser erythropus* Linn.) Barnacle Goose. Is said to have been shot at Quincy, Mass., by Dr. S. Cabot Jr., (Proc. B. S. N. H., vol. III, p. 136.) Prof. Baird says, "its occurrence in North America is very doubtful, resting only on very insufficient evidence. (P. R. R. Ex. and Surv., vol. IX, p. 768.)

57. *Nettion crecca* Kaup. English Teal. Accidental from Europe. Has been taken in the State. (Dr. H. Bryant, Proc. B. S. N. H., vol. V, p. 195.)

58. *Spatula clypeata* Boie. Spoonbill. Shoveller. Not uncommon. Chiefly seen in spring and fall.

59. *Mareca penelope* Bon. European Widgeon. Has been taken at several points along the eastern coast of the United States, and has been found apparently breeding on Long Island, (Dr. T. M. Brewer, Proc. B. S. N. H., vol. VI, p. 419) where it has been repeatedly found. One has been taken in this State. (E. A. Samuels.)

60. *Fulix marila* Baird. Scaup Duck. Black-headed Duck. "Blue Bill." Not common. Found chiefly in spring and autumn; occasionally in winter.

61. *Fulix affinis* Baird. Little Black-headed Duck. Spring and fall. Not common.

62. *Fulix collaris* Baird. Ring-necked Duck. Spring and autumn. Rare.

63. *Aythya americana* Bon. Red-headed Duck. "Redhead." Autumn and winter. Not very common. Abundant in the markets of Boston in winter, but, like the Canvass-backs, are brought from the bays and rivers of the Middle States.

64. *Histrionicus torquatus* Bonap. Harlequin Duck. Winter visitant. Not common.

65. *Camptolæmus labradorius* Gray. Labrador Duck. Rare winter visitant.

66. *Pelionetta perspicillata* Kaup. Surf Duck. Common in fall and spring, and some remain through the winter.

67. *Oidemia americana* Swain. Scoter. Autumn and winter. Not uncommon; often abundant.

68. *Somateria mollissima* Leach. Eider Duck. Not uncommon in winter.

69. *Somateria spectabilis* Leach. King Eider. Rare visitant in winter.

The Smew (*Mergellus albellus* Selby) Mr. E. A. Samuels attributes to this State, having seen a specimen which he was told was taken in Massachusetts Bay.

The American Pelican (*Pelecanus erythrorhyncus* Gmelin) has recently been taken at Calais, Me., (G. A. Boardman, Proc. B. S. N. H., vol. IX, p. 130) and, according to DeKay, was formerly numerous on the Hudson and other rivers and lakes of New York. It probably occurs as a chance visitor in this State.

70. *Sula bassana* Ross. Gannet. Occasional on the sea coast in fall and winter.

71. *Graculus carbo* Gray. Common Cormorant. "Shag." Common near the coast in fall and winter.

72. *Graculus dilophus* Gray. Double-crested Cormorant. Not uncommon near the coast in winter.

73. *Procellaria glacialis* Linn. Fulmar Petrel. Spring and autumn visitant.

74. *Thalassidroma Wilsonii* Bon. Wilson's Petrel. Not uncommon off the coast. Have seen specimens taken near Chelsea Beach.

75. *Thalassidroma pelagica* Bon. Mother Cary's Chicken. Rare, off the coast, as far south as Provincetown. (E. A. Samuels.)

76. *Puffinus major* Bon. Greater Shearwater. Not common. Off the coast in winter.

77. *Puffinus fuliginosus* Strick. Sooty Shearwater. Coast in autumn and winter. Not common.

78. *Puffinus anglorum* Temn. Mank's Shearwater. Rare, off the coast in winter.

79. *Stercorarius pomarinus* Temn. Pomarine Jager. Massachusetts Bay. Rare in winter.

80. *Stercorarius parasiticus* Temn. Arctic Jager. Massachusetts Bay in winter. Not common.

81. *Stercorarius cepphus* Lawr. Buffon's Skau. Rare. Has been taken near Boston.

82. *Larus marinus* Linn. Black-backed Gull. Not a common winter visitant.

83. *Larus delawarensis* Ord. Ring-billed Gull. Not very uncommon near the coast in winter.

84. *Larus leucopterus* Fabr. White-winged Gull. Rare winter visitant.

85. *Chrœcocephalus atricilla* Lawr. Laughing Gull. Winter. Not common.

86. *Rissa tridactyla* Bon. Kittiwake Gull. Very common about the islands in Massachusetts Bay in autumn and winter.

The Fork-tailed Gull (*Xema Sabinii* Bonap.) may occur on our coast as an occasional visitor.

87. *Sterna aranea* Wils. Marsh Tern. Rare summer visitor. (E. A. Samuels, Agr. Mass. 1863, Sec's Rep. App., p. xxix.)

88. *Sterna fuliginosa Gm.* Sooty Tern. Rare summer visitor. (E. A. Samuels, Agr. Mass., 1863, Secy's Rep., App., p. xxix.) Mr. Samuels informs me that it breeds on Muskegat Island, near Martha's Vineyard.

89. *Sterna hirundo* Linn. Wilson's Tern. Common in summer, breeding on the rocky islands in the Bay.

90. *Sterna macroura* Naum. Arctic Tern. Common. Chiefly a winter visitant. Sometimes breeds.

91. *Sterna paradisea* Brünn. Roseate Tern. Very rare; perhaps merely accidental; from the south in summer. Several instances known of its capture in the State. (Chelsea Beach, Nuttall, Man. Orn. vol. II, p. 278.—Beverly, Mass., 1847, S. Cabot Jr., Proc. B. S. N. H. vol. II, p. 248.—Have seen a specimen taken off our coast a few years since.

92. *Sterna frenata* Gambel. Least Tern. Spring and

autumn visitant. Not common. Occasional in summer.

Other species of *Sterna* undoubtedly occur as rare visitors off our coast, as the Caspian Tern (*Sterna caspia* Pall.) in winter, from the north; and possibly, at the same season, Trudeau's Tern (*Sterna Trudeauii* Aud.) as a very rare, or accidental species.

93. *Hydrochelidon fissipes* Gray. (*Hydrochelidon plumbea* Wils.) Short-tailed Tern. Not common. Have seen specimens taken near Chelsea Beach.

94. *Colymbus arcticus* Linn. Black-throated Diver. rare autumn and winter visitor.

95. *Utamania torda* Leach. Razor-billed Auk. Not uncommon on the coast in autumn and winter.

96. *Mormon arctica* Ill. Arctic Puffin. Not uncommon in winter.

97. *Uria grylle* Lath. Black Guillemot. Not very uncommon winter visitant.

98. *Cataractes troile* Bryant. Foolish Guillemot. Murre. Not uncommon in winter, and perhaps a few breed.

99. *Cataractes ringvia* Bryant. Murre. Common in winter.

100. *Cataractes lomvia* Bryant. (*Uria arra* Pall.) Thick-billed Guillemot. Murre. Rather common in winter.

101. *Mergulus alle* Vieill. Little Auk. Sea Dove. Rather rare winter visitant. Has been taken on the Connecticut, at Greenfield, Mass., in one instance.

The birds found in Massachusetts may be conveniently grouped into the following classes: I. Species that regularly breed in the State. II. Resident species. III. Winter visitants. IV. Spring and autumn visitants. V. Summer visitants. VI. Accidental or irregular visitants.

I. *Species that regularly breed in the State.*

Those marked with a star, though repeatedly found breeding in some localities, breed very sparingly, and not generally over the State. Some others are common in some parts of the State, but are unknown or very rare in most parts. Several others have been known to breed, but apparently only accidentally, as in the case of *Chrysomitris pinus*, *Spizella monticola*, and a few others. A few not in the list may occasionally breed.

1. Falco anatum *Bonap.**
2. Tinnunculus sparverius *Vieill.**
3. Accipiter Cooperii *Bonap.*
4. " fuscus *Bonap.*
5. Buteo borealis *Vieill.*
6. " lineatus *Jard.*
7. " pennsylvanicus *Bon.*
8. Circus hudsonius *Vieill.*
9. Haliætus leucocephalus *Sav.**
10. Pandion carolinensis *Bon.**
11. Bubo virginianus *Bonap.*
12. Scops asio *Bonap.*
13. Otus americanus *Bonap.*
14. Brachyotus Cassinii *Brewer.*
15. Syrnium nebulosum *Gray.*
16. Nyctale acadica *Bonap.*
17. Coccygus americanus *Bon.*
18. " erythrophthalmus *Bon.*
19. Picus villosus *Linn.*
20. " pubescens *Linn.*
21. Sphyropicus varius *Baird.*
22. Hylotomus pileatus *Baird.**
23. Melanerpes erythrocephalus *Sw.**
24. Colaptes auratus *Swain.*
25. Trochilus colubris *Linn.*
26. Chætura pelasgia *Steph.*
27. Antrostomus vociferus *Bon.*
28. Chordeiles popetue *Baird.*
29. Ceryle alcyon *Boie.*
30. Tyrannus carolinensis *Bd.*
31. Myiarchus crinitus *Cab.**
32. Sayornis fuscus *Baird.*
33. Contopus borealis *Baird.*
34. " virens *Cab.*
35. Empidonax Traillii *Baird.*
36. " minimus *Bd.*
37. " acadicus *Bd.*
38. Turdus mustelinus *Gmelin.*
39. " Pallassi *Cabanis.**
40. " fuscescens *Steph.*
41. " migratorius *Linn.*
42. Sialia sialis *Baird.*
43. Mniotilta varia *Vieill.*
44. Parula americana *Bon.*
45. Geothlypis trichas *Cab.*
46. Icteria viridis *Bon.**
47. Helmitherus vermivorus *Bon.**
48. Helminthophaga ruficapilla *Bd.*
49. Siurus aurocapillus *Swain.*
50. " noveboracensis *Nutt.*
51. Dendroica virens *Baird.*
52. Dandroica canadensis *Bd.**
53. " Blackburniæ *Bd.*
54. " pinus *Baird.*
55. " æstiva *Bd.*
56. " discolor *Bd.*
57. Euthlypis canadensis *Cab.**
58. Setophaga ruticilla *Sw.*
59. Pyranga rubra *Vieill.*
60. Hirundo horreorum *Barton*
61. " lunifrons *Say.*
62. " bicolor *Vieill.*
63. Cotyle riparia *Boie.*
64. Progne purpurea *Boie.*
65. Ampelis cedrorum *Baird.*
66. Vireo olivaceus *Vieillot.*
67. " gilvus *Bonap.*
68. " noveboracensis *Bon.*
69. " flavifrons *Vieill.*
70. Mimus polyglottus *Boie.**
71. Galeoscoptes carolinensis *Cab.*
72. Harporynchus rufus *Cabanis.*
73. Cistothorus palustris *Cab.*
74. " stellaris *Cab.*
75. Troglodytes aedon *Vieill.*
76. Certhia americana *Bon.*
77. Sitta carolinensis *Gmel.*
78. Parus atricapillus *Linn.*
79. Carpodacus purpureus *Gray.*
80. Astrigalinus tristis *Cabanis.*
81. Passerculus savanna *Bon.*
82. Poœcetes gramineus *Baird.*
83. Coturniculus passerinus *Bon.*
84. " Henslowii *Bon.**
85. Ammodromus caudacutus *Sw.*
86. " maritimus *Sw.*
87. Junco hyemalis *Sclater.**
88. Spizella pusilla *Bonap.*
89. " socialis *Bonap.*
90. Melospiza melodia *Baird.*
91. Helospiza palustris *Baird.*
92. Guiraca ludoviciana *Swain.*
93. Cyanospiza cyanea *Baird.*
94. Pipilo erythrophthalmus *Vieill.*
95. Dolichonyx oryzivora *Swain.*
96. Molothrus pecoris *Swainson.*
97. Agelæus phœniceus *Vieill.*
98. Sturnella magna *Swain.*
99. Icterus spurius *Bon.**
100. " Baltimore *Bon.*
101. Quiscalus versicolor *Vieill.*
102. Corvus americanus *Aud.*

103. Cyanura cristata *Sw.*
104. Ectopistes migratoria *Sw.*
105. Zenædura carolinensis *Bon.*
106. Cupidonia cupido *Baird.**
107. Bonassa umbellus *Steph.*
108. Ortyx virginianus *Bonap.*
109. Ardea herodias *Linn.*
110. Ardetta exilis *Gray.*
111. Botaurus lentiginosus *Steph.*
112. Butorides virescens *Bon.*
113. Nyctiardea Gardeni *Baird.*
114. Oxyechus vociferus *Reich.*
115. Ægialeus melodus *Reich.*
116. Squartarola helvetica *Cuv.*
117. Philohela minor *Gray.*
118. Gallinago Wilsonii *Bonap.*
119. Symphemia semipalmata *Hartl.*
120. Tringoides macularius *Gray.*
121. Bartramia laticauda *Less.*
122. Rallus virginianus *Linn.*
123. Porzana carolina *Vieill.*
124. Fulica americana *Gmel.**
125. Anas obscura *Gmel.**
126. Mergus americanus *Cass.**
127. " serrator *Linn.**
128. Aix sponsa *Boie.*
129. Thalassidroma Leachii *Temm.*
130. Sterna hirundo *Linn.*
131. " macroura *Naum.*

II. *Resident Species.*

Of a few species more properly to be regarded as spring and autumn or summer visitors, a few individuals are sometimes found in winter, as of *Ceryle alcyon*, *Turdus migratorius*, *Melospiza melodia*, &c., but since the majority are migratory, they are not placed in the list of resident species.

1. Falco anatum *Bon.*
2. Tinnunculus sparverius *Vieill.*
3. Buteo borealis *Vieill.*
4. Circus hudsonius *Vieill*
5. Haliætus leucocephalus *Savigny.*
6. Bubo virginianus *Bonap.*
7. Scops asio *Bonap.*
8. Otus americanus *Bon.*
9. Brachyotus Cassinii *Brew.*
10. Syrnium nebulosum *Gray.*
11. Nyctale acadica *Bonap.*
12. Picus villosus *Linn.*
13. " pubescens *Linn.*
14. Hylotomus pileatus *Baird.*
15. Certhia americana *Bon.*
16. Sitta carolinensis *Gmel.*
17. Parus atricapillus *Linn.*
18. Astrigalinus tristis *Cab.*
19. Corvus americanus *Aud.*
20. Cyanura cristata *Swain.*
21. Cupidonia cupido *Baird.*
22. Bonasa umbellus *Steph.*
23. Ortyx virginiana *Bonap.*
24. Fulica americana *Gmel.*
25. Anas obscura *Linn.*
26. Mergus americana *Cass.*
27. " serrator *Linn.*
28. Colymbus torquatus *Brunn.*

III. *Winter Visitants.*

Those species marked with a star are occasional or irregular visitors, but some of them sometimes occur in great abundance. A few individuals are often found in winter of some of those species properly to be considered as spring and autumn visitants, and as such are placed in the next list below.

1. Astur atricapillus *Bon.*
2. Archibuteo lagopus *Gray.*
3. " Sancti-Johannis *Gr.*
4. Aquila canadensis *Cass.**
5. Syrnium cinereum *Aud.**
6. Nyctale Richardsonii *Bon.**
7. Nyctea nivea *Gray.*
8. Picoides arcticus *Gray.**
9. Regulus satrapa *Licht.*
10. Ampelis garrulus *Linn.**
11. Collyrio borealis *Baird.*
12. Troglodytes hyemalis *Vieill.*
13. Sitta canadensis *Linn.*
14. Parus hudsonicus *Forst.**
15. Eremophila cornuta *Boie.*
16. Pinicola canadensis *Cab.**
17. Chrysomitris pinus *Bonap.*
18. Curvirostra americana *Wils.*
19. " leucoptera *Wils.**
20. Ægiothus linaria *Cab.**

21. Plectrophanes nivalis *Meyer.*
22. Centrophanes lapponicus *Kaup.**
23. Spizella monticolor *Baird.*
24. Tetrao canadensis *Linn.**
25. Arquatella maritima *Baird.*
26. Anser hyperboreus *Pallas.*
27. " Gambellii *Hartl.*
28. Dafila acuta *Jenyns.*
29. Bucephala americana *Baird.*
30. " albeola *Baird.*
31. Histrionicus torquatus *Bon.*
32. Camptolæmus labridorius *Gr.*
33. Melanetta velvetina *Baird.*
34. Pelionetta perspicillata *Kaup.*
35. Oidemia americana *Swain.*
36. Somateria mollissima *Leach.*
37. " spectabilis *Leach.*
38. Erismatura rubida *Bonap.*
39. Lophodytes cucullatus *Reich.*
40. Sula bassana *Briss.*
41. Graculus carbo *Gray.*
42. " dilophus *Gray.*
43. Puffinus major *Faber.*
44. " fuliginosus *Strick.*
45. Puffinus anglorum *Temm.*
46. Stercorarius pomarinus *Temm.*
47. " parasiticus *Temm.*
48. " cepphus *Ross.*
49. Larus leucopterus *Faber.**
50. " marinus *Linn.*
51. " Smithsonianus *Coues.*
52. " delawarensis *Ord.*
53. Chrœcocephalus atricilla *Lawr.*
54. " Philadelphia *Lawr.*
55. Rissa tridactyla *Bonap.*
56. Sterna macroura *Naum.*
57. Colymbus septentrionalis *Linn.*
58. Podiceps Holbollii *Reinh.*
59. " cristatus *Lath.*
60. " cornutus *Lath.*
61. Utamania torda *Leach.*
62. Mormon arctica *Illiger.*
63. Uria grylle *Latham.*
64. Cataractes troile *Bryant.*
65. " ringvia *Bry.*
66. " lomvia *Bry.**
67. Mergulus alle *Vieill.*

IV. *Spring and Autumn Visitants.*

Of some species properly regarded as spring and autumn visitants, a few individuals remain through the winter, in sheltered situations, or through the summer, now and then breeding. Those of which some remain in winter are marked with this * ; those in summer, with this †. There may be a few other species of this character not thus marked, as *Empidonax flaviventris*, *Vireo solitarius*, &c., that should be.

1. Hypotryorchis columbarius *Gr.*
2. Pandion carolinensis *Bon.*
3. Empidonax flaviventris *Baird.*
4. Turdus Pallassi *Cab.*†
5. { Turdus Swainsonii *Cab.* / " aliciæ *Baird.* }
6. Regulus calendula *Licht.*
7. Anthus ludovicianus *Licht.*
8. Geothlypis Philadelphia *Bd.*
9. Oporornis agilis *Baird*
10. Helmitherus Swainsonii *Bon.*
11. Helminthophaga pinus *Baird.*
12. " chrysoptera *Bd.*
13. " celata *Baird.*
14. " peregrina *Baird.*
15. Dendroica coronata *Gray.*
16. " castanea *Baird.*
17. " striata *Bd.*
18. " maculosa *Bd.*
19. " tigrina *Bd.*
20. Dendroica palmarum *Bd.*
21. Wilsonia pusilla *Bon.*
22. Euthlypis canadensis *Cab.*†
23. Vireo solitarius *Vieill.*
24. Zonotrichia leucophrys *Swain.*
25. " albicollis *Bon.*
26. Junco hyemalis *Sclat.**†
27. Helospiza Lincolnii *Baird.*
28. Passerella illiaca *Sw.*
29. Scolecophagus ferrugineus *Sw.*
30. Charadrius virginicus *Borck.*
31. Ægialeus semipalmatus *Reich.*
32. Strepsilas interpres *Illig.*
33. Phalaropus Wilsonii *Cab.*
34. " hyperboreus *Temn.*
35. " fulicarius *Bonap.*
36. Macroramphus griseus *Leach.*
37. Tringa canutus *Linn.*
38. Ancylochilus subarquata *Kaup.*
39. Pelidna americana *Coues.*

40. Actodromas maculata *Cass.*
41. " pusillus *Coues.*
42. " Bonapartii *Cass.*
43. Calidris arenaria *Illig.*
44. Ereunetes pusilla *Cass.*
45. Micropalama himantopus *Bd.*
46. Gambetta melanoleuca *Bon.*
47. " flavipes *Bon.*
48. Rhyacophilus solitarius *Bon.*
49. Tringites rufescens *Cab.*
50. Limosa fedoa *Ord.*
51. " hudsonica *Swain.*
52. Numenius longirostris *Wils.*
53. " hudsonicus *Lath.*
54. " borealis *Lath.*
55. Porzana noveboracencis——?
56. Bernicla canadensis *Boie.*
57. " Hutchinsii *Bon.*
58. Bernicla brenta *Steph.*
59. Anas boschas *Linn.*
60. Nettion carolinensis *Bd.*
61. Querquedula discors *Steph.*
62. Spatula clypeata *Boie.*
63. Chaulelasmus streperus *Gr.*
64. Mareca americanca *Steph.*
65. Fulix marila *Baird.*
66. " affinis *Bd.*
67. " collaris *Bd.*
68. Aythya americana *Bonap*
69. " vallisneria *Bon.*
70. Harelda glacialis *Leach.*
71. Procellaria glacialis *Linn.*
72. Sterna frenata *Gambel.*
73. Hydrochelidon fissipes *Gray*
74. Podilymbus podiceps *Lawr.*

V. *Summer Visitants.*

Of some species of which the greater part are merely summer visitants a few individuals remain in winter, but not enough to entitle the species to be considered resident, and are marked thus * ; those of which the greater part pass north to breed, thus † ; those of which but few reach us in summer from the south, thus ‡.

1. Accipiter Cooperii *Bon.*
2. " fuscus *Bon.*
3. Buteo lineatus *Jard.**
4. " pennsylvanicus *Bon.*
5. Coccygus americanus *Bon.*
6. " erythrophthalmus *Bp.*
7. Sphyropicus varius *Baird.*
8. Melanerpes erythrocephalus *Sw.*
9. Colaptes auratus *Swain.*
10. Trochilus colubris *Linn.*
11. Chætura pelasgia *Steph.*
12. Antrostomus vociferus *Bon.*
13. Chordeiles popetue *Baird.*
14. Ceryle alcyon *Boie.**
12. Tyrannus carolinensis *Bd.*
16. Myiarchus crinitus *Cab.*‡
17. Sayornis fuscus *Baird.*
18. Contopus borealis *Baird.*
19. " virens *Cab.*
20. Empidonax Traillii *Bd.*
21. " minimus *Bd.*
22. " acadicus *Bd.*
23. Turdus mustelinus *Gmel.*
24. " fuscescens *Steph.*
25. " migratorius *Linn.**
26. Sialia sialis *Baird.*
27. Mniotilta varia *Vieill.*†
28. Parula americana *Bon.*
29. Geothlypis trichas *Cab.*
30. Icteria viridis *Bon.*‡
31. Helmitherus vermivorus *Bon.*
32. Helminthophaga ruficapilla *Bd.*
33. Siurus aurocapillus *Sw.*
34. " noveboracensis *Nutt.*†
35. Dendroica virens *Bd.*
36. " canadensis *Bd.*†
37. " Blackburniæ *Bd.*†
38. " pinus *Bd.*
39. " pennsylvanicus *Bd.*
40. " æstiva *Bd.*
41. " discolor *Bd.*
42. Wilsonia minuta *Bon.*
43. Setophaga ruticilla *Sw.*
44. Pyranga rubra *Vieill.*
45. Hirundo horreorum *Bart.*
46. " lunifrons *Say.*
47. " bicolor *Vieill.*
48. Cotyle riparia *Boie.*
49. Progne purpurea *Boie.*
50. Ampelis cedrorum *Baird.*
51. Vireo olivaceus *Vieill.*
52. " gilvus *Bonap.*
53. " noveboracensis *Bon.*
54. " flavifrons *Vieill.*

55. Mimus polyglottus *Boie.*‡
56. Galeoscoptes carolinensis *Cab.*
57. Harporhyncus rufus *Cab.*
58. Cistothorus palustris *Cab.*
59. " stellaris *Cab.*
60. Troglodytes ædon *Vieill.*
61. Carpodacus purpureus *Gr.**†
62. Passerculus savanna *Bon.*
63. Poœcetes gramineus *Bd.*
64. Coturniculus passerinus *Bon.*
65. " Henslowii *Bon.*
66. Ammodromus maritimus *Sw.*
67. " caudacutus *Sw.*
68. Spizella pusilla *Bon.*
69 " socialis *Bon.*
70. Melospiza melodia *Baird.*
71. Helospiza palustris *Bd.*
72. Euspiza americana *Bon.*
73. Guiraca ludovicana *Sw.*
74. Cyanospiza cyanea *Baird.*
75. Pipilo erythophthalmus *Vieill.*
76. Dolichonyx oryzivorus *Sw.*
77. Molothrus pecoris *Sw.*
78. Agelæus phœniceus *Vieill.*
79. Sturnella magna *Swain.**
80. Icterus spurius *Bon.*‡
81. Icterus Baltimore *Daud.*
82. Quiscalus versicolor *Vieill.*
83. Ectopistes migratorius *Sw.*
84. Zenædura carolinensis *Bon.*
85. Ardea herodias *Linn.*
86. Ardetta exilis *Gray.*
87. Botaurus lentiginosus *Steph.*
88. Butorides virescens *Bon.*
89. Nyctiardea Gardenii *Bd.*
90. Oxyechus vociferus *Reich.*
91. Ochthodromus Wilsonius *Reich.*‡
92. Ægialeus melodus *Cab.*
93. Squartarola helvetica *Cur.*
94. Hæmatopus palliatus *Temm.*
95. Philohela minor *Gray.*
96. Gallinago Wilsonii *Bon.*
97. Symphemia semipalmata *Hartl.*
98. Tringoides macularius *Gray.*
99. Bartramia lacticauda *Less.*
100. Rallus virginianus *Linn.*
101. Porzana carolina *Vieill.*
102. Aix sponsa *Boie.*
103. Sterna hirundo *Linn.*
104. " paradisea *Brunn.*‡
105. " aranea *Wils.*
106. " fuliginosa *Gm.*‡

VI. *Accidental and Irregular Visitors.*

The following species are known in this State merely as rare chance visitors, a few only excepted, from the common habitat of their respective species. The others are very irregular in their visits. There are many other species so extremely rare that there are but few known instances of their capture in the State ; but from what is known of their distribution we are not to regard them in the light of chance visitors.

1. Cathartes atratus *Lesson.*
2. " aura *Ill.*
3. Falco candicans *Gmelin.*
4. Syrnium cinereum *Aud.*
5. Nyctale Richardsonii *Bon.*
6. Picoides arcticus *Gray.*
7. Centurus carolinus *Bon.*
8. Icteria viridis *Bon.*
9. Pyranga æstiva *Vieill.*
10. Ampelis garrulus *Linn.*
11. Mimus polyglottus *Boie.*
12. Parus hudsonicus *Forster.*
13. Curvirostra leucoptera *Wils.*
14. Ægiothus linaria *Cab.*
15. Centrophanes lapponicus *Kaup.*
16. Chondestes grammaca *Bon.*
17. Helospiza Lincolnii *Baird.*
18. Euspiza americana *Bon.*
19. Cardinalis virginianus *Bon.*
20. Quiscalus major *Vieill.*
21. Corvus carnivorus *Bartram.*
22. " ossifragus *Wils.*
23. Tetrao canadensis *Linn.*
24. Garzetta candidissima *Bon.*
25. Herodias egretta *Gray.*
26. Florida cærulea *Baird.*
27. Ibis Ordii *Bon.*
28. Ochthodromus Wilsonius *Reich.*
29. Rallus crepitans *Gm.*
30. Gallinula galeata *Bon.*
31. " martinica *Lath.*
32. Bernicla leucopsis *Linn.*
33. Nettion crecca *Kaup.*
34. Mareca penelope *Bon.*
35. Sterna paradisea *Brunn.*
36. " fuliginosa *Gm.*

Summary.

Number of species found at Springfield	195
" " " " in the State	296
" " " that breed in the State	131
" " resident species	28
Winter visitants	67
Spring and Autumn visitants	75
Summer visitants	106
Chance visitors	35

Springfield, April, 1864.

Supplemental Notes. While the preceding paper has been passing through the press the following facts have been ascertained:

A specimen of the Snowy Owl (*Nyctea nivea* Gray) was taken in Springfield, the present year, about May 20th. Another instance of its capture here late in May has occurred within a few years. It has been found here repeatedly in November, and consequently spends at least half the year here.

A specimen of the Yellow-billed Cuckoo (*Coccygus americanus* Bonap.) was killed here in May this year, by Mr. B. Hosford, who informs me that he obtained another specimen here several years since. The capture of only three specimens of this species at Springfield has as yet come to my knowledge.

That the Hermit Thrush (*Turdus Pallasi* Cab.) does occasionally breed at Springfield I am now convinced, having seen a specimen shot here in July, but the instances appear to be extremely rare

In the preceding list of the Birds of Springfield, the Prairie Warbler (*Dendroica discolor* Baird) is mentioned as rare, and as not breeding at Springfield. I find it not uncommon in sandy fields, growing up thinly to pitch pines, the present summer, where it is breeding quite plentifully. It was not uncommon in June to hear several males singing at a time.

A pair of Yellow-breasted Chats (*Icteria viridis* Bonap.) are breeding here the present season. Noticed another pair in Ludlow, Mass., about June 3d, which were probably also breeding. Have seen a specimen taken in Berkshire county, in the breeding season. Only straggling pairs of this species, however, reach Massachusetts.

The following species of Hawks, though extremely rare in winter, should probably be properly included in the above list of "Resident Species:" *Hypotriorchis columbarius* Gr., *Accipiter Cooperii* Bon., *A. fuscus* Bon., *Buteo lineatus* Jard., and *B. pennsylvanicus* Bon.

July, 1864.

V. *Notes on the Habits of some species of Humble Bees.* By F. W. Putnam.

(Communicated October 22, 1863.)

During the summer of 1862, while in Warwick, Mass., my attention was called to the Humble Bees by finding three nests of *Bombus fervidus* Fabr. and *B. vagans* Smith. These nests were formed of the deserted nests of mice, one under a barn in an old stump of a tree, the other two under piles of stones in a field. One of the nests of *B. fervidus* I kept in a box for some time, and watched the actions of the bees, but as I then neglected to make full notes, and as my first observations were confirmed by later ones, I allude to them here only to introduce an incident which has relation to the duration of life of the various kinds which always compose the communities of the Humble Bees. Upon leaving Warwick I left my valise, in which was a nest of bees, at the depot. Two months afterwards, in November, it was brought to me, when upon examining the nest several large queen bees were found in a lively condition, while the males, small females and workers were all dead. When the valise was left at the depot there was but one queen in the nest. This incident proves that the queens are not only late in leaving the cells, but that they are capable of enduring cold which is fatal to the other bees. In the summer of 1863 while at Bridport, Vt., on the borders of Lake Champlain, I was so successful, in my entomological excursions, as to find as many as twenty-five or thirty colonies of bees, and to collect fifteen complete nests. These were of the following species: *Bombus fervidus* Fabr., *B. ternarius* Say, *B. separatus* Cresson and *B. virginicus* Fabr. As the general economy of these four species is the same, my observations may be considered as made upon one community, preceded however by the following special statements in regard to the several species.

BOMBUS TERNARIUS. Two nests collected: one under an old stump in a deserted mouse nest; the other, in September, under the clapboards of a house, about eight feet from the ground. Upon removing the boards, a large bunch of sheep's wool was found, evidently collected by rats, as there was a quantity of nut shells, with the under jaw and other bones of a rat among the wool. In the centre of the wool the bees had their cells. By etherizing the bees twenty-eight specimens were collected, which, as it was after dark, when the bees are generally at home, I have reason to believe were nearly all that belonged to the nest. There were thirty-five cells containing young, and thirty that were filled with honey, having their tops covered with wax. This is the only instance of my finding *the honey cells closed over*. There were also a number of bunches of pollen in which there were no eggs.

This species is not so common as B. fervidus and is far more savage in its disposition. I was informed by Mr. Brigham Rockwood, that he had noticed that this species never takes possession of the nests of mice *(Arvicola)* which are found so plentifully among the grass, but always chooses a place under cover of boards or stumps.

BOMBUS FERVIDUS. This is the most common species at Bridport, and is of quite a gentle disposition, allowing its nest to be disturbed for some time before it makes any show of resistance, merely exhibiting its uneasiness by buzzing. The communities of this species are found in old mice nests, both under stumps and boards; and also among the grass in the nests of the common field mice *(Arvicola riparia)*. They also occupy the forsaken nests of the house mice, as in one instance a colony was found under the flooring of a shed, in a nest made of bits of paper, rags, &c. This was the largest community collected, consisting of about seventy adult bees, one hundred and fifty cells containing young, and two hundred young larvæ, in various stages of growth, in the pollen masses, besides fifty cells filled with honey. This nest was found on the 23d of July. July 28th a nest was discovered in which there was a single queen bee and five or six large queen cells still soft and recently finished.

July 8th. Two queens were seen fighting upon the outside of a nest. So firmly were they united that they did not part until placed in alcohol, although pushed about for some time. They were both of the same species, but one might have been an invader, as I have found upon placing a strange queen, in a nest, that the rightful sovereign immediately commenced battle and in a short time expelled the intruder.

One community kept under glass on a window, with free ingress and egress, continued working, until, on a very hot day, the young became baked in their cells, by the heat of the sun. Then the old ones left and did not return.

Aug. 6th. A nest was brought home and the cells, containing young, placed apart from all old bees for the purpose of ascertaining if the young bee cuts its own way out of its cell. The cells were all of large size. In about half an hour a queen bee had come out and was seen walking over the other cells. She was immediately removed and the other cells were examined, but no signs of their having been cut could be seen. In the evening a slit was noticed in one of the cells and the young bee was seen at work cutting with its jaws. In a short time it made an opening in the cell large enough for it to push its head through. It then commenced cutting on each side, from the slit, above and below; now and then withdrawing its head and resting. Then it tried to force its way through the opening, but finding this was not large enough it cut a little more. The bee evidently did not wish to work more than was necessary, for it often tried to force its way out. At each attempt it made but a small enlargement of the orifice; but, after spending half an hour in alternate work and rest, it succeeded in freeing itself from its prison. Then it stood, for a short time, on the sides of the cell, moving its wings, after which it commenced walking over the other cells. This was a queen bee. Aug. 8th, another bee came out in the same way. Aug. 10th, two. Aug. 14th, one. Aug. 15th, another, which was the last in the cells. They were queens and all quite light colored when just from the cells.

These facts prove that the young cut their own way out of the cells. In another nest a young bee was seen to come from the cell while the old bees were present, which did not concern themselves about the matter further than to give a few passing glances and to cut off some jagged pieces of the cell. As soon however as the young bee was out of the cell, one or two old bees trimmed the edges of the cell and removed a few fragments from the inside.

BOMBUS SEPARATUS. Several colonies of this were found under old stumps and in other situations similar to those in which the nests of B. fervidus were found. This species is nearly as ferocious, on being disturbed, as B. ternarius.

BOMBUS VIRGINICUS. A single nest of this species was found under an old stump in an orchard. On the 27th of August three males were captured while flying under a large tree on which they frequently alighted. So much did these bees resemble large flies in their actions, that at first I mistook them for those insects. Male Humble Bees are often seen flying in this manner under trees. Are they not the drones which have left or been driven from the nest?

Let us now notice the life of a colony in its different stages. In the spring, the queen bee, having left her old home, may be seen roaming about in search of a new one, which she soon finds in some such place as previously described. She immediately collects a small amount of pollen mixed with honey, and in this deposits from seven to fourteen eggs, gradually adding to the pollen mass until the first brood is hatched. She does not wait, however, for one brood to be hatched before laying the eggs of another, but, as soon as food enough has been collected, she lays the eggs for a second. The eggs are laid, in contact with each other, in one cavity of the mass of pollen, with a part of which they are slightly covered. They are very soon developed; in fact the lines are nowhere distinctly drawn, between the egg and the larva, the larva and pupa, and again between the latter and the imago; a perfect series, showing this gradual transformation of the young to the imago, can be found in almost every nest.

As soon as the larvæ are capable of motion and commence feeding they eat the pollen by which they are surrounded, and gradually separating, push their way in various directions. Eating as they move and increasing in size quite rapidly, they soon make large cavities in the pollen mass. When they have attained their full size they spin a silken wall about them, which is strengthened by the old bees covering it with a thin layer of wax, which soon becomes hard and tough, thus forming a cell. The larvæ now gradually attain the pupa stage and remain inactive until their full developement. They then cut their way out and are ready to assume their duties as workers, small females, males or queens according to their individual formation.

It is apparent that the irregular disposition of the cells is due to their being constructed so peculiarly by the larvæ. After the first brood, composed of workers, has come forth, the queen bee devotes her time principally to her duties at home, the workers supplying the colony with honey and pollen. As the queen continues prolific, more workers are added and the nest is rapidly enlarged.

About the middle of summer, eggs are deposited which produce both small females and males, and it is supposed by some observers that it is from the union of these, at the last of the season, that the eggs are laid from which the queens are developed: but there seems some reason to doubt this, as a new nest, previously mentioned, was found on the last of July occupied only by a queen and queen larvæ. It is true, however, that all eggs, laid after the last of July, produce the large females, or queens, and, the males being still in the nest, it is presumed that the queens are impregnated at this time, as, on the approach of cold weather all, except the queens, of which there are several in each nest, die.

The efforts of my friend Mr. Rockwood to procure nests for me during the winter have as yet been unsuccessful, those which he had marked for removal having been destroyed by mice.

It is desirable to ascertain whether the queens remain torpid during cold weather and what use is made of the pollen and honey stored during the last of summer and in the fall, which perhaps is food for the queens during the mild weather in spring before plants are in blossom.

But little wax is made by the Humble Bees, as it is only used for covering the cocoons of the larvæ, for thinly lining the nest on the inside, strengthening the old cells which are used for honey pots, and occasionally covering these pots, and propping up the old cells.

During some years Humble Bees are very numerous. This is generally the case when a dry and early spring is followed by a summer producing a good crop of clover. After such a season, if the following spring be favorable, nests are very abundant.

Though very similar to those made by Reaumer, over a hundred years ago, it will be noticed that my observations differ, in several particulars, from those made by some European naturalists who have written on the Bombi.

Some observers have stated that the eggs of the Humble Bee are deposited in cells, partly filled with pollen, which are enlarged by the workers as the young increase in size, and that the old bees, cutting holes in the cells, feed the young until they are fully developed when they relieve them from their prisons. This is quite contrary to the results of my observations in New England.

At present I cannot believe that the peculiarity of food, or the structure of the cells, produces a difference of developement in Humble Bees, for the larvæ, as has been previously stated, were seen to make their own cells from the pollen paste, while the old bees were quite indiscriminate in selecting the plants from which they procured both pollen and honey.

Is it not more natural to believe, as has been suggested to me by Professor Wyman, that the difference in the developement of the eggs is owing to their being laid at various times after impregnation? Thus, if I am right in supposing that the queens are impregnated by the males late in the summer, the eggs laid soon after produce the large queen larvæ: the next set of eggs, laid in the spring, produce the workers, or undeveloped females, while from those deposited still later, male bees are principally developed.

This opinion seems to be corroborated by the state of the nest, previously noticed, found on the 28th of July, which had been recently commenced and contained only *queen* cells, the parent queen being obliged, by her recent impregnation, to lay only such eggs as were adapted to the season. As no first brood of workers, or second one of males and small females, had existed in this nest, the eggs producing the queen larvæ must have been laid by the large female or queen, found in the nest, and not by a small female.

The fact, that our species of Humble Bees take possession of the nest of mice and rats, accounts for the large number of *mites* found in most nests.

Three parasites are common in the nests of our New England Humble Bees. They are, a small beetle of the genus *Byturus* only known thus far in the imago state; a moth of the genus *Nephopteryx*: the larvæ of which is quite abundant in most nests, and a dipterous insect which is often found in the larval state.

It is singular that in all the nests, which I collected, not a single specimen of *Apathus* was found by Mr. Packard, though this parasitic bee is generally supposed to be quite common in the nests of Bombi.

Additional Notes, August 3, 1864. A nest of *Bombus pennsylvanicus* was found at Upton, Me., on the sixth of last June, in which there was but a single queen bee with seven cells of the smallest size, containing larvæ, and several eggs in a mass of pollen.

A queen of *B. pennsylvanicus* was taken, on July 20th, under leaves in a wood.

Professor A. E. Verrill found a queen Humble Bee in a torpid state under leaves, before the snow was off the ground in the spring of 1863.

IX. *Classification of Polyps: (Extract condensed from a Synopsis of the Polypi of the North Pacific Exploring Expedition, under Captains Ringgold and Rodgers, U. S. N.).* By A. E. VERRILL.

(Communicated February 29, 1865.)

The report upon the collection made by Dr. William Stimpson, naturalist to the expedition, having been much delayed, the following tabular view of the classification adopted is here presented, with the hope that, if imperfect like every other, it may, nevertheless, afford some aid in illustrating the natural affinities of these humble forms.

Although in a communication read before a Zoölogical Club at Cambridge, Jan. 1862, I attempted to demonstrate the existence of the three natural orders among polyps, I refrained from presenting this view in a paper published last year, in order that I might make further investigations upon the subject before finally publishing it.

CLASS CNIDARIA OR POLYPI.

ORDER I. MADREPORARIA.

Polyps simple or compound with embryonic or rudimentary basal or abactinal region, which has no special function unless for vegetative attachment while young. Actinal area well developed, form broadly expanded, having a tendency in the higher groups to become narrowed towards the mouth. Tentacles simple, conical. Dermal tissues, and usually the radiating lamellæ, depositing solid coral; the radiating plates being between the lamellæ, are, therefore, ambulacral and appear to originate from the surfaces of the lamellæ and the connective tissues extending across the ambulacral chambers and filling them from below. Interambulacral spaces distinct.

Suborder I. Stauracea (*Madreporaria rugosa**).

Coral simple, or compound by budding; chiefly *epidermal* and *endothecal*; *septa* apparently in multiples of four, sometimes wanting. Type embryonic, like a young Astrea or Fungia.

Families,—*Stauridæ, Cyathophyllidæ, Cyathaxonidæ, Cystiphyllidæ.*

Suborder II. Fungacea.

Polyps either simple or compound by marginal or disk budding, rarely by fissiparity. Tentacles numerous, in multiples of six, imperfectly developed, scattered on the actinal surface, usually short and lobe-like. Upper part of polyps scarcely exsert. Coral broad and low, growth mostly centrifugal, tissue chiefly septal; walls imperfectly developed, often perforate, subordinate, usually forming the basal attachment.

Families,—*Cyclolitidæ, Lophoseridæ, Fungidæ, Merulinidæ.*

Suborder III. Astreacea.

Polyps mostly compound, either by fissiparity or various modes of budding. Tentacles usually well developed, long, subcylindrical, limited in number, in multiples of six

* This group is placed here with considerable hesitation and principally on account of the close resemblance in structure to the young of the succeeding and higher groups, when they first begin to form a coral, which then consists of a ring of epitheca or epidermal deposit with a few, imperfect, rugose septa radiating from the centre. If the number four be a constant feature of the arrangement of their septa, it is possible that they may be entitled to rank as a separate order of Polyps. To this opinion Prof. J. D. Dana inclines. Prof. Agassiz unites the group with Hydroid Acalephs on account of their resemblance, in some features, to the *Tabulata*. It seems to me, however, that the absence of transverse plates in *Cyathaxonidæ* and *Cystiphyllidæ* and the perfection of the vertical septa in *Stauridæ*, *Cyathaxonidæ* and some of the *Cyathophyllidæ*, together with their general structure, shows them to be more closely allied to the *Fungacea* and *Astreacea*, of which they may be considered embryonic types, while at the same time the group is a synthetic one, having analogies with nearly all the higher groups of Polyps and also in some respects, with Hydroids.

encircling the disk. Coral mural, septal and endothecal; growth vertical and centrifugal, producing turbinated forms which are often elongated.

Families,—*Lithophyllidæ*, *Mœandrinidæ*, *Eusmillidæ*, *Caryophyllidæ*, *Stylinidæ*, *Astreinæ*, *Oculinidæ*, *Stylophoridæ*.

Suborder IV. Madreporacea (*Madreporaria perforata*).

Tentacles in definite numbers, twelve or more, well developed, encircling the narrowed disk, therefore nearer the mouth; polyps with the upper portion much exsert, flexile; growth chiefly vertical; coral mural and septal, porous. Polyps compound by budding, sometimes simple.

Families,—*Eupsammidæ*, *Gemmiporidæ*, *Poritidæ*, *Madreporidæ*.

ORDER II. ACTINARIA.

Polyps with well developed, often highly specialized, basal or abactinal region. Walls well developed, tentacles longer, more concentrated around the mouth, which is also, usually, if not always, furnished with special tentacular lobes or folds. Ambulacral spaces always open, destitute of connecting tissues and solid deposits.

Suborder I. Zoanthacea.

Polyps encrusting, adherent, budding from mural expansions; tentacles simple, short, at edge of disk.

Families,—*Zoanthidæ*, *Bergidæ*.

Suborder II. Antipathacea.

Polyps connected by a cœnenchyma, secreting a solid sclerobase or coral axis. Tentacles few, six to twenty-four, simple, conical.

Families,—*Antipathidæ, Gerardidæ.*

Suborder III. Actinacea.

Polyps free, capable of locomotion, with a highly specialized, muscular base or abactinal area. Tentacles well organized, either simple or branched, varying from ten to many hundreds, often with accessory organs arising from the same spheromeres, such as inner tentacles, verrucæ, complicated or simple branchial lobes, cinclidæ, eye-spherules, suckers, etc. Mouth with special lobes or folds. Most of the species are simple, a few are compound by fissiparity, many abnormally bud from the wall near the base, a few secrete from the base a horn-like deposit similar to the axis of Antipathes.

Families,—*Actinidæ, Thalassianthidæ, Minyidæ, Ilyanthidæ, Cerianthidæ.*

ORDER III. ALCYONARIA.

Polyps with well developed actinal, mural and abactinal regions, compound by budding. Tentacles eight, pinnately lobed, long, encircling a narrow disk. No inter-ambulacral spaces. Ambulacral ones open and wide.

Suborder I. Alcyonacea.

Polyps turbinate at base, budding in various ways, encrusting, adherent to foreign bodies by the cœnenchyma.

Families,—*Alcyonidæ, Xenidæ, Cornularidæ, Tubiporidæ.*

Suborder II. Gorgonacea.

Polyps cylindrical, short, connected by a cœnenchyma, secreting a central supporting axis.

Families,—*Gorgonidæ, Plexauridæ, Primnoidæ, Gorgonellidæ, Isidæ, Corallidæ, Briaridæ.*

Suborder III. Pennatualacea.

Polyps forming free, moving colonies, the composite basal portion with locomotive functions and special cavities, with or without a solid free axis.

Families,—*Pennatulidæ, Pavonaridæ, Veretillidæ, Renillidæ.*

Among the most interesting species in this collection the following may be mentioned:

Stephanoseris lamellosa Verrill.

Coral low, subcylindrical, with a broad base, which completely covers small univalve shells with the exception of the opening; wall rudimentary; septa in four cycles, the primaries much the largest with subentire rounded tops; columella well developed, papillose, costæ prominent, unequal.

Loo Choo Islands. Dr. Wm. Stimpson.

Heterocyathus alternata Verrill.

A low species with very unequal septa and costæ, the primary septa very prominent. Encrusts and covers small univalve shells.

Gaspar straits. Capt. John Rodgers.

Balanophyllia capensis Verrill.

A species about half an inch high, broadly attached, slightly turbinated, with an epitheca rising within a line of the margin. Calicle deep, broadly oval. Septa in four cycles, the principal ones much exsert, vertical, narrowed at top, those of the fourth cycle joining the columella in pairs. Color of the living polyp bright orange.

Cape of Good Hope. Dr. Wm. Stimpson.

EUPSAMMIA STIMPSONII Verrill.

Coral free, elongated, turbinated, blunt at base. Calicle oval, deep; columella well developed, septa broad, the principal ones with entire inner edges, rounded. Length an inch or more; breadth of cell .30

Interesting as a living representative of a genus hitherto known only in the fossil state.

North China Sea. Dr. Wm. Stimpson.

METRIDIUM FIMBRIATUM Verrill.

A species closely allied to *M. marginatum* of this coast, but apparently more elongated, with longer and more slender tentacles which are almost hair-like. Disk within the tentacles narrow. "Color pale orange, translucent, body punctate with dark brown, mouth deep orange."

San Francisco, Cal. Dr. Wm. Stimpson.

PHELLIA COLLARIS Verrill.

Edwardsia collaris Stimpson, Proc. Philad. Acad. Nat. Science, May and June, 1855.

A species remarkable for its great size compared with previously known species from Europe.

Hong Kong, China. Dr. Wm. Stimpson.

PHELLIA CLAVATA Verrill.

Edwardsia clavata Stimpson, l. c. 1855.

A species even larger than the last.

Near Ousima, Japan. Dr. Wm. Stimpson.

AMMONACTIS nov. gen.

Column elongated, subcylindrical, with well developed basal disk, covered, as in *Phellia*, with a persistent epidermis extending to near the summit, naked above; but differs in having a lobe-like tubercle below each tentacle, distinct from the margin. Tentacles long and numerous.

AMMONACTIS RUBRICOLLUM Verrill.

Edwardsia rubricollum Stimpson, l. c. 1855.

Hong Kong, China. Dr. Wm. Stimpson.

HALOCAMPA BREVICORNIS Verrill.

Edwardsia brevicornis Stimpson, l. c. 1855.

Hong Kong, China. Dr. Wm. Stimpson.

HALOCAMPA CAPENSIS Verrill.

Body elongated, tentacles twenty, blunt; ambulacra subpapillose. Six tentacles have their inner bases dark brown; body pale reddish with dots and patches of flake white; inner side of tentacles flake white.

Cape of Good Hope, 12 fathoms, sand. Dr. William Stimpson.

CERIANTHUS ORIENTALIS Verrill.

A large species similar to *C. americana* nobis. Body elongated, in a tube of mud. Tentacles long and slender. Color of body deep reddish brown, outer tentacles translucent, yellowish and white, pale brown on their inner sides, greenish at base; inner ones purplish brown or sometimes grass green.

At low water mark, Hong Kong, China. Dr. William Stimpson.

NEPHTHYA THYRSOIDEA Verrill.

Polyps forming thyrsiform bunches of closely clustered branchlets, three inches high and two broad. Color wine-yellow or light brown, with a dark purplish tinge below the tentacles; tentacles nearly white; spicula forming elevated, transverse lines of silvery white on the stalks.

Cape of Good Hope, 20 fathoms, rocks. Dr. Wm. Stimpson.

TELESTO RAMICULOSA Verrill.

Cornularia aurantiaca Stimpson, l. c. 1855, non. *T. aurantiaca* Lamx.

Hong Kong, 10 fathoms, shelly bottom. Dr. Wm. Stimpson.

PARISIS LAXA Verrill.

Coral forming openly reticulate fronds; papillæ numerous rounded, on all sides of the branches; cœnenchyma minutely villous in alcohol. Calcareous joints shorter and internodes longer than in *P. fruticosa* nobis.

Hong Kong. Dr. Wm. Stimpson.

ACANTHOGORGIA COCCINEA Verrill.

Nepthya coccinea Stimpson, l. c. 1855.

Hong Kong, 10 fathoms, on shells. Dr. Wm. Stimpson.

VERETILLUM STIMPSONII Verrill.

A large species six or eight inches long, the upper portion enlarged, more than half the entire length. Polyps much exsert, upwards of an inch long; tentacles very long. Axis thick, short, fusiform, a third of an inch long. Base white, somewhat striated; body light cream-color; polyps transparent, bluish white at the bases of the tentacles.

Hong Kong, 6—10 fathoms, mud. Dr. Wm. Stimpson.

VERETILLUM BACULATUM Verrill.

Club-shaped, the base about a third of the length. Polyps scattered, not numerous. Axis small, fusiform, about half an inch long in a specimen three inches long.

Sea of Ochotsk, off Siberia. L. M. Squires.

KOPHOBELEMNON CLAVATUM Verrill.

Veretillum clavatum Stimpson, l. c. 1855.

Polyps more numerous and crowded than in *K. Burgeri* Herkl. which it resembles; body more claviform, naked dorsal space very narrow.

Hong Kong, 6 fathoms, mud. Dr. Wm Stimpson.

XII. *Synopsis of the Polyps and Corals of the North Pacific Exploring Expedition, under Commodore C. Ringgold and Captain John Rodgers, U. S. N., from* 1853 *to* 1856. *Collected by Dr. Wm. Stimpson, naturalist to the Expedition. With Descriptions of some additional Species from the West Coast of North America.* By A. E. Verrill.

Part II, Alcyonaria. With two Plates.

[Communicated February 29, 1865.]

The specimens upon which the following descriptions are based were mainly collected by Dr. Wm. Stimpson while acting as naturalist to the expedition.

They were for the most part preserved in alcohol, and many are accompanied by notes and drawings of the soft parts, which have been reproduced in the plates. In most instances I have given the descriptions of the colors of perishable parts, as well as notes on the mode of occurrence, in Dr. Stimpson's own words.

Descriptions of a few species in the collection of the Smithsonian Institution and the Yale College museum, from the Pacific Coast of America, have been added, for the sake of making thepaper more complete.

SUBORDER, PENNATULACEA.

Family, Pennatulidæ.

Pteromorpha expansa Verrill.

Plate 5, *figure* 1.

The pinnate portion is broad ovate, abruptly rounded below; peduncle, or basal portion, thick, swollen, a little less than half the entire length. Pinnæ crowded, about thirty-two on each side, long and wide, somewhat thickened, angular, the naked posterior margin somewhat concave, the anterior rounded and supporting numerous small polyps, and strengthened with sharp spines, which are often in clusters of two or three. The outer half of the sides of the pinnæ as well as their anterior edges, are covered by small polyp-cells; basal half of the lower surface densely covered by small papillæ. Axis strong, pointed at the ends; interior cavity of the base small. Length of a large specimen in alcohol 6 inches, breadth across pinnæ 3.5, length of peduncle 2.75.

"Color (in life) white, bases of the polyps dirty white, on the stalk there are a few scattered blackish spots.

It lives with the stalk immersed in the mud like *Renilla;* undulating, moving contractions are often seen in the stalk, resembling those of a *Holothuria.*"

Bays opposite Hong Kong, China. Common in 6 fathoms, mud, April, 1854. Dr. Wm. Stimpson.

LEIOPTILUM Gray.

This genus is most nearly allied to *Pennatula*, but differs in having soft, fleshy pinnæ, with even borders and no apparent spicula. The polyps are in two or more rows along the edges of the pinnæ. The peduncle is enlarged into a conspicuous, contractile bulb. The axis is very slender, quadrangular, and extends only through the middle portion of the body. The rudimentary individuals on the back are developed in the form of conspicuous papillæ.

LEIOPTILUM UNDULATUM Verrill, nov. sp.

Basal portion smooth, pointed at the end, swelling into a large bulb just below the pinnæ. Posterior part of the body, except along a narrow median band, covered with large verruciform rudimentary polyps, forming rounded papillæ, some of which are a tenth of an inch in diameter. Pinnæ large, very broad and rounded, with narrow bases, the edges thrown into undulations or frills. Polyps rather large, arranged in three alternating rows along the edges of the pinnæ. Axis very slender, about two inches long, extending from about an inch above the basal end to about the middle of the pinnate portion. The naked base, of a specimen 4.25 inches long, is 1.75; the largest pinnæ .75 long and 1.12 wide. This specimen has twenty-five pinnæ on each side. Pinnacati Bay, Cal. Mr. Stone. (Coll. Smithsonian Inst.)

PTILOSARCUS.

This section of the genus *Sarcoptilus* Gray seems to be sufficiently distinct from the original type of that group to rank as a separate genus.

The form is thick, club-shaped; the pinnæ numerous, crowded, with thickened edges on which the polyps are arranged in several rows, each cell surrounded by prominent, spine-like spicula. The back of the body, except along a narrow median space, is covered by two broad bands of rudimentary polyps, appearing like crowded granulations. The basal portion is thick and bulbous, with two large interior cavities, one of which extends along the anterior surface, communicating with the pinnæ, the other along the dorsal portion.

The axis is long, fusiform, tapering to very slender points, which are curved (in preserved specimens) into a loop at each end. Connected with the lower part of the axis are very strong thickened muscles, which pass obliquely upward and outward to the wall-tissues, while higher up, a little above the lowest pinnæ, other shorter ones are attached, which pass obliquely downward to the wall.

PTILOSARCUS GURNEYI.

Sarcoptilus (*Ptilosarcus*) *Gurneyi* J. E. Gray, Ann. and Mag. Nat. Hist., Vol. 5, p. 23, pl. III, f. 2, 1860; ? *Pennatula tenua* Gabb. Proc. Cal. Acad. Nat. Sciences, II. page 166. 1862.

Basal portion about one half the whole length, thick, bulbous, very muscular, the surface strongly sulcated in contraction. Pinnæ smooth on the sides, broad, rounded, nearly semicircular with a broad base, the posterior edge extending beyond the base as a rounded lobe: the edge is thickened and covered by the polyps arranged in four rows, each cell armed with two sharp spinules. Along the back are two broad bands of very small papillæ or granuliform, rudimentary polyps.

Length of a large alcoholic specimen, having fifty-two pinnæ on each side, 10 inches; greatest breadth 2; length of pinnæ .80, breadth 1.50; length of naked base 4.75, diameter 1.25.

Puget Sound, Wash. Terr. Dr. C. B. Kennerly, Dr. G. Suckley. (Coll. Smithsonian Inst.)

FAMILY, PAVONARIDÆ Dana, restricted.

VIRGULARIA PUSILLA Verrill, nov. sp.

Plate 5, *figure* 2.

Very small and slender, the pinnæ extending nearly to the base, which is rounded and bulbous; pinnæ of the upper portion surrounding the stalk on all sides except the back, which is naked; below they are separated also by a narrow anterior space, but the pinnæ of the opposite sides appear to coalesce anteriorly higher up, producing a subverticillate arrangement. The middle whorls are separated about .1 of an inch; polyps small, twelve to fourteen in the median whorls, somewhat crowded; tentacles slender, elongated, with slender, rather distant, lateral lobes along nearly their whole extent. Length 1.75 inches; diameter at the middle .12.

"Bays opposite Hong Kong, China, in 6 fathoms, mud. April, 1854. Color pale orange or dirty red." Dr. Wm. Stimpson.

The only specimen in the collection is probably young.

FAMILY, VERETILLIDÆ Gray, emended.

VERETILLUM STIMPSONI Verrill, these Proceedings, p. 152, April, 1865.

Plate 5, *figures* 3, 3*a*.

Polypiferous portion of surface thick, swollen, somewhat fusiform, broadest below the middle, the surface granulous; basal portion less than a third of the whole length, bulbous, smooth, and very contractile; polyps rather distantly scattered, arranged somewhat in quincunx; between them are numerous minute papillæ or rudimentary polyps; in expansion the polyps are much exsert, with very slender elongated tentacles, bordered with rather distant, elongated, slender lobes in a single row on each side, commencing close to their bases; axis short, thick, fusiform, situated just below the commencement of the polypiferous part.

Whole length of the largest specimen in alcohol 3.5 inches; naked part 1; diameter where broadest 1; length of axis .35. When living, length 6.5 inches; breadth 1.75; polyps about .75, exsert.

"Hong Kong Harbor, China, in 6 to 10 fathoms, mud, March 1855; also in 24 fathoms, shelly sand, China Sea, 23° N. lat. April, 1855. Body whitish cream-colored; polyps transparent with an opaque digestive tube, bluish white about the bases of the tentacles; base white, somewhat longitudinally striated." Dr. Wm. Stimpson.

VERETILLUM BACULATUM Verrill, these Proc. p. 152, April, 1865.

Small, clavate, broadest near the upper end, which is obtusely rounded; polypiferous portion about one half the whole length; naked basal portion elongated, pointed below, in one specimen, with a distinct terminal pore; axis small, fusiform, less than one half an inch long; polyps much smaller and more numerous than in the preceding.

Length of the only specimen obtained 2 inches; diameter .3.

Sea of Ochotsk, in 25 fathoms, taken Aug. 1855, by U. S. Steamer "John Hancock," Capt. Stevens; preserved by L. M. Squires.

KOPHOBELEMNON CLAVATUM Verrill, l. c. page 152.

Veretillum clavatum Stimpson, Proc. Phil. Acad. Nat. Sciences, Vol. 7, p. 375, June, 1855.

Plate 5, *figures* 4, 4*a*, 4*b*.

Polyps large, the tentacles long and slender with oblong lateral lobes; surface of the body between the polyps, irregularly papillose, variegated, punctate with orange and spotted with brown; basal portion white, with a pointed extremity. Length 2 inches.

Bay opposite Hong Kong, in 6 fathoms, mud, April, 1854. Dr. Wm. Stimpson.

This species is more claviform and has more crowded polyps than K. Burgeri Herklotz. The naked dorsal space is scarcely apparent, owing to the crowding of the polyps towards it upon each side.

SUBORDER, GORGONACEA.

Family, Gorgonidæ.

Gorgonia venosa Valenciennes.

Off Madeira, in 25 fathoms, rocks. Dr. Wm. Stimpson.

Leptogorgia cuspidata Verrill, nov. sp.

Corallum broad, subflabelliform, irregularly branching nearly in a plane, the principal branches arising near the base divide above in an irregularly dichotomous manner, forming a rather thick fasciculate clump. Branchlets thick, rigid, nearly straight, tapering to a point. Cells numerous, rather large, rounded, covering the surface of the branchlets except along a narrow median space on each side. Grooves rarely distinct except near the base. Color deep purple; cells yellow; axis black.

Cape St. Lucas, Cal. J. Xantus. (Coll. Smithsonianl Institution).

Family, Plexauridæ Gray.

Plexaura friabilis Lamouroux.

Cape of Good Hope. Dr. Wm. Stimpson.

I refer with some doubt to this species, specimens of a large dichotomous *Plexaura* with long, upright, cylindrical branches, the terminal ones often undivided for a foot or more, and about .3 of an inch in diameter, tapering but little at the ends. The cells are often a little prominent and evenly crowded; the axis dark brown, scarcely compressed, even at the axils; the cœnenchyma very spiculose and friable.

It resembles in form and general appearance *P. crassa* (Gorgonia vermiculata Lk.) of the West Indies.

Lophogorgia palma Edw. and Haime.

Gorgonia palma Pallas, 1766. *Gorgonia flammea* Ellis and Solander, 1786.

False Bay, Cape of Good Hope. Not rare in 20 fathoms, rocks, Oct. 1853. Dr. Wm. Stimpson.

One specimen differs from the ordinary form in having a large, very compressed trunk, with the long, subdigitate branches much more flattened than usual; color, in alcohol, light gray.

LISSOGORGIA Verrill.

Proc. Boston Soc. Nat. History, 1864.

In this genus the cœnenchyma is very thin, friable and highly spiculose throughout, the spicula conspicuous at the surface and covering the verruciform cells, which are eight-lobed in contraction. Tentacles strengthened at the base by large spicula, which often radiate within the dried cells. Axis horn-like, smooth, usually without visible striations. Type, *L. flabellum* (Antipathes flabellum Auth.)

LISSOGORGIA FLEXUOSA Verrill, nov. sp.

Corallum much branched, subflabelliform, the branches irregularly pinnate, branchlets slender, divaricate, often coalescing; axis soft and flexible, dark brown; cœnenchyma thin, membranous, filled with large fusiform spicula, visible at the surface; polyp cells rather large, rounded, verruciform, covered by numerous elongated and pointed, imbricated spicula. Color, in alcohol, grayish white.

Hong Kong. Dr. Wm. Stimpson.

MURICEA SINENSIS Verrill, nov. sp.

Plate 5, figures 5, 5a.

Corallum irregularly dichotomous with elongated, subclavate branchlets; polyp cells verruciform, rather large, somewhat prominent, irregularly crowded, surface granulose with crowded spicula; external portion of the cœnenchyma hard and coriaceous, rather thick, the surface thickly covered by small oblong spicula; tentacles strengthened by numerous red spicula. Axis, in alcohol, very soft and flexible near the ends, slender, dark brownish below; color of the cœnenchyma deep red. Height 8 inches, diameter of branchlets .15.

Hong Kong. Dr. Wm. Stimpson.

MURICEA ? DIVARICATA Verrill, nov. sp.

Plate 5, *figures* 6, 6*a*.

Corallum low, much branched somewhat in a plane, branchlets slender, elongated, divaricate; covered by the very prominent irregularly crowded, sometimes secund, polyp cells; these are mostly .2 of an inch high and spread abruptly at right angles to the branches, and are somewhat claviform, the summits being enlarged. The cœnenchyma is very thin and filled with large, thickened spicula, conspicuous at the surface, producing a granulated appearance; the polyp cells are thickly covered by more elongated, fusiform spicula, which are irregularly arranged, interlaced and conspicuous at the surface, converging at the summit of the cells, which are eight-rayed in contraction. Color, in alcohol, light ash gray, axis light fuscous, soft and flexible. Height three inches.

Hong Kong. Dr. Wm. Stimpson.

ACANTHOGORGIA COCCINEA Verrill, these Proceedings, p. 152.

Nephthya coccinea Stimpson, Proc. Phil. Acad. Nat. Sci. June, 1855, Vol. 7, p. 375.

Plate 6, *figures* 7, 7*a*.

All the specimens observed of this species consist of simple clavate stalks, rising from broadly expanded bases, which, like the stalk, are densely covered with large, open polyp cells, irregularly crowded over the whole surface; the cells are surrounded by numerous, deep red, prominent, imbricated, spines, their outward ends long and sharp, but irregularly branched at their bases, forming thus a cluster of short, secondary spines; axis light brown, slender and flexible. Height of the largest specimens 2 inches, diameter .2. Below each tentacle, imbedded in the external membrane, are two rows of linear, crimson spicula, converging towards each other so as to form a series of V-shaped markings with the apex towards the ends of the tentacles.

Hong Kong, China, in 10 fathoms, attached to dead shells. "The specimens when contracted look like the fruit of the Sumac of New England (*Rhus typhina L.*) Color bright red, that of the tentacles and spicula deepest; polyp bodies hyaline, yellowish flesh color." Dr. Wm. Stimpson.

FAMILY, PRIMNOIDÆ M. Edwards, emended.

PRIMNOA COMPRESSA Verrill, nov. sp.

Corallum much branched in a plane, flabelliform, consisting of several large branches arising from near the base, which give off, alternately from each side, numerous, long, slender, acute branchlets, which rise at a very acute angle with the main branches and are often again subdivided in the same way; branches and branchlets strongly compressed in the plane of the branches, delicately striated, stony, near the base dark brown, the branchlets yellowish white, their tips setaceous. Height of largest specimen 24 inches; diameter of largest branches .25. Cœnenchyma and polyps not observed.

Aleutian Islands. Capt. Gibson.

FAMILY, GORGONELLIDÆ Valenciennes.

JUNCELLA LÆVIS Verrill, nov. sp.

Corallum tall, simple, subcylindrical, rather slender, diminishing in size both at the summit and near the base, where the polyps become obsolete; cells appressed, scarcely prominent, arranged in two broad bands, leaving a narrow, median, naked space on each side, along which there is a well marked groove; they are placed alternately at a distance of about .2 of an inch, in about six vertical rows on each side, producing a quincunx arrangement; axis slender, cylindrical, calcareous, white, surrounded by about sixteen longitudinal tubes, two of which are larger and correspond with the lateral grooves, the others to the rows of polyps. Length of the single specimen, imperfect at each end, 20 inches; greatest diameter .25. Color yellowish brown, in alcohol.

Hong Kong, China. Dr. Wm. Stimpson.

Family, Isidæ.

Parisis laxa Verrill, these Proceedings, p. 152.

Corallum flabelliform, loosely branched, openly reticulated, only a few of the branches coalescent; branchlets spreading nearly at right angles, somewhat elongated, curved, obtuse at the ends; papillæ rather large, irregularly crowded; cœnenchmya thin, roughened by the points of minute spicula; axis slender, consisting of white calcareous joints alternating with shorter dark brown ones of the same thickness, but softer; color, in alcohol, light gray. Height of a small specimen 3 inches; width 3; diameter of branchlets .20.

Off Hong Kong, China, in 15 fathoms, shelly gravel, May, 1854. Color, in life, bright light blue. Dr. Wm. Stimpson.

Mopsella japonica Verrill, nov. sp.

Low, spreading, dichotomous, branching nearly in a plane; branches slender, diverging at an angle of about 45,° obtuse at the ends; cells rounded, papilliform, rather large, crowded. Color, of all the specimens observed, vermillion with yellow polyps.

Simoda, Japan. Dr. Wm. Stimpson.

This species is most nearly allied to *M. coccinea* (Isis coccinea Ellis and Sol.), but the branchlets do not coalesce as in that species, and spread much less abruptly. The cells, also, are considerably larger.

SUBORDER, ALCYONACEA.

Family, Alcyonidæ.

Alcyonium rubiforme Dana.

Lobularia rubiformis Ehrenberg.

Arctic Ocean in 35 fathoms. Capt. J. Rodgers. West Coast of Behrings Straits in the Laminarian Zone. Dr. Wm. Stimpson.

ALCYONIUM, SP.

A specimen badly preserved, and too imperfect for identification.

Hong Kong. Dr. Wm. Stimpson.

ALCYONIUM?

The corallum consists of rounded, glomerate clusters of large, verruciform polyps, striated at the tops. Color, in alcohol, bright red.

Sea of Ochotsk. L. M. Squires.

SARCOPHYTON AGARICUM Verrill.

Alcyonium agaricum Stimpson, l. c. page 375.

This species forms mushroom-shaped disks, which are circular, convex, with entire, revolute margins, and supported on a central pedicel about one-third as broad as the disk.

The polyps cover only the upper surface, are rather large, three-tenths of an inch long and an eighth of an inch distant, the surface between them being covered with minute dots. Upper surface of the disk bluish gray, polyps lighter with still paler tentacles; lower surface and pedicel dark cream colored. Diameter of disk 1.5 to 2 inches; of pedicel .5.

Kagosima Bay, Japan, not uncommon in 10 fathoms, sand, January, 1855. Dr. Wm. Stimpson.

NEPHTHYA AURANTIACA Verrill, nov. sp.

Corallum thyrsoid in form, consisting of a stout naked pedicel, divided above into several short, thick branches, which are covered by small, glomerate clusters of crowded polyps; cells very small, verruciform, much crowded; their bases covered with closely imbricated, red spicula; the bases of the tentacles with golden yellow ones. Height 2 inches; diameter of pedicel .3. Color of pedicel and branches pale pink; polyp cells bright red at base, yellow at summit.

China Sea in 23° N. lat. Dredged in 28 fathoms, shelly gravel. Color pale reddish gray; tentacles light yellow Dr. Wm. Stimpson.

NEPHTHYA THYRSOIDEA Verrill, l. c. p. 151.

Plate 6, *figures* 8, 8*a*, 8*b*.

Corallum thyrsoid, consisting of a pyramidal head of compound, glomerate clusters of polyp cells, supported by a short, thick pedicel. The short branches arise from all sides of the main trunk and spread abruptly, dividing at once into numerous small rounded lobes, which are densely covered by the crowded polyps; cells larger than in the preceding, less thickly covered by the spicula, which are yellowish gray and quite small. Height of the largest specimen, 3 inches, diameter 2, diameter of pedicel .5, length of naked part .75.

False Bay, Cape of Good Hope. "Taken commonly in small clusters, rarely in large ones, in 20 fathoms, rocks, Oct. 1853. Color wine-yellow or light brown; polyps dark purplish just under the tentacles; the latter palish, nearly white; stalks with irregular, transverse, elevated, silvery lines of spicula." Dr. Wm. Stimpson.

SPONGODES GIGANTEA Verrill, Bulletin of the Museum of Comparative Zoölogy, p. 40. Jan., 1864.

Large, paniculately branched; principal branches few and large, covered on all parts by short, thick, glomerate branchlets, which are themselves divided into numerous clusters or small heads of polyps; the polyps are small, not crowded, most of them armed with a bundle of long, white, prominent spines, some with smaller single ones; bases of the tentacles filled with numerous red spicula; the trunk is very open, cavernous, the walls membranous, filled with slender, white spicula; the base divided into root-like expansions. Height 12 inches or more; diameter of trunk, near the base 3; of principal branches 2. Color of trunk, in alcohol, brownish gray; polyps dark red, with conspicuous white spines.

Hong Kong, China, on rocks in 1 fathom, April, 1854. Dr. Wm. Stimpson.

SPONGODES CAPITATA Verrill, Bulletin of the Museum of Comparative Zoölogy, p. 40. Jan., 1864.

Trunk short and thick, dividing rapidly in a dichotomous manner, forming a broad rounded clump. Branchlets much subdivided, corymbose, the polyps all terminal, small, verruciform in contraction, in rounded clusters of forty or fifty; with these are often intermingled little groups of three or four individuals supported by slender, white spines, one of which is longer than the rest, and considerably exsert, supporting one of the cells on its side. The tentacles and polyp walls are also strengthened by slender white spicula.

Height of a large specimen 4 inches, breadth 5, diameter of polyps about .05.

Hong Kong, China. Dr. Wm. Stimpson.

SPONGODES GRACILIS Verrill, nov. sp.

Trunk slender, arborescently branched, the branchlets lax, a little elongated, furcate, the polyps mostly arising singly along their sides in a secund manner, not crowded, scarcely clustered, small but quite prominent, supported by several slender spines, the walls and tentacles strengthened by numerous very slender fusiform spicula of a bright red color. The trunk and branches are open, membranous, diaphanous, filled with long, slender, curved, fusiform spicula of a white color.

General color light pink. Height 2 inches; diameter of trunk .25; of polyps .02.

Loo Choo Islands. Dr. Wm. Stimpson.

FAMILY, CORNULARIDÆ.

ANTHELIA LINEATA Stimpson, l. c. p. 375.

Plate 6, *figures* 9, 9*a*, 9*b*.

Polyps elongated, tapering somewhat towards the disk. Tentacles nearly half as long as the body, slender, tapering, with a single series of oblong, somewhat irregular papillæ, those near the base of the tentacles shorter and nearly obsolete. Body pale brownish with eight, longitudinal, lead-colored stripes, tentacles bright blue. Length of body about an inch.

Hong Kong. "Abundant on rocks at low-water mark, which it covers with a light bluish scum." Dr. Wm. Stimpson.

TELESTO RAMICULOSA Verrill.

Cornularia aurantiaca Stimpson, l. c. p. 375. (non T. aurantiaca Lamx.)

Plate 6, *figures* 10, 10*a*, 10*b*.

Corallum irregularly branching, the branches and corallites straight, subcylindrical (clavate when young) marked with numerous, fine, longitudinal sulcations. The upper portion of the polyps projecting considerably beyond the firmer tubes in expansion, pellucid, somewhat constricted at the junction with the tubes. Tentacles long and rather broad, with a single series of elongated lateral lobes, which are themselves tuberculated. Color pale orange, polyps transparent with a few linear spicula on the sides, stomach crimson. Height 2 inches.

Hong Kong. Dredged sparingly in 10 fathoms, shelly bottom. Dr. Wm. Stimpson.

TELESTO? NODOSA Verrill, nov. sp.

In this species the stalks usually rise half an inch or more and then divide at once into a cluster of twelve to fifteen slender corallites or branchlets, which diverge from one point. By continued growth another leading polyp arises from the cluster and after a short distance produces another similar cluster, thus forming an elongated stalk, densely ramulous along the sides, except where the intervals between the clusters occur, and even there scattered cells often appear. Corallites slender, turbinate, a third of an inch long, the walls very thin, encircled by numerous elevated rings, and finely striated longitudinally.

Height of the largest specimen about three inches.

Loo Choo Islands, in pools at low water. Dr. Wm. Stimpson.

Dry specimens only of this species are in the collection, and therefore its true characters are somewhat uncertain. In many respects, and especially in the transverse rings around the tubes, it resembles some species of *Tubularia*.

SARCODICTYON.

A species of this genus occurs creeping over dead shells. The polyps in alcohol are mostly contracted and form scattered verrucæ about a line in diameter and rather more in height, connected by slender, fleshy stolons.

Hong Kong, China. Dr. Wm. Stimpson.

FAMILY, TUBIPORIDÆ.

TUBIPORA RUBEOLA? Quoy and Gay.

Bonin Islands. Dr. Wm. Stimpson.

I refer to this species, with much doubt, a fragment of *Tubipora* of very light red color, having the transverse plates about .25 of an inch apart; the tubes .1 in diameter. It is too imperfect to be determined satisfactorily, if, indeed, it be possible to determine species of this genus from the dry corals.

EXPLANATION OF THE PLATES.

PLATE V.

Figure 1. PTEROMORPHA EXPANSA Verrill; a polyp enlarged.

Figure 2. VIRGULARIA PUSILLA Verrill; a polyp much enlarged.

Figure 3. VERETILLUM STIMPSONI Verrill; a polyp somewhat enlarged; 3*a*, structure of the surface between the polyps.

Figure 4. KOPHOBELEMNON CLAVATUM Verrill; natural size, with the polyps expanded, front view; 4*a*, a polyp much enlarged, showing outline of stomach, below which are, apparently, clusters of eggs; 4*b*, a tentacle much magnified.

Figure 5. MURICEA SINENSIS Verrill; a branchlet, natural size; 5*a*, one of the contracted cells magnified to show the spicula. This and the next were drawn from nature by Mr. E. S. Morse.

Figure 6. MURICEA DIVARICATA Verrill; portion of a branch, natural size; 6*a*, a contracted cell much magnified, showing spicula

Plate VI.

Figure 7. Acanthogorgia coccinea Verrill; a polyp much enlarged, showing spiniform spicula at the base and small spicula below the tentacles; 7*a*, view of a polyp from above.

Figure 8. Nephthya thyrsoidea Verrill; a cluster of polyps from a branch, slightly magnified; 8*a*, a polyp partially contracted, much enlarged to show the spicula; 8*b*, one of the spicula greatly magnified.

Figure 9. Anthelia lineata Stimpson; cluster of polyps natural size; 9*a*, a polyp enlarged; 9*b*, a tentacle magnified.

Figure 10. Telesto ramiculosa Verrill; a group of polyps in contraction, natural size; 10*a*, an expanded polyp much enlarged; 10*b*, a contracted polyp somewhat magnified.

Note. Part I of this paper, including a Synopsis of the Classification of Polyps, herein adopted, was published under a somewhat different title in the fifth number of these Proceedings, page 145. Part III, embracing descriptions of the *Actinaria*, will appear as soon as the necessary plates can be completed.

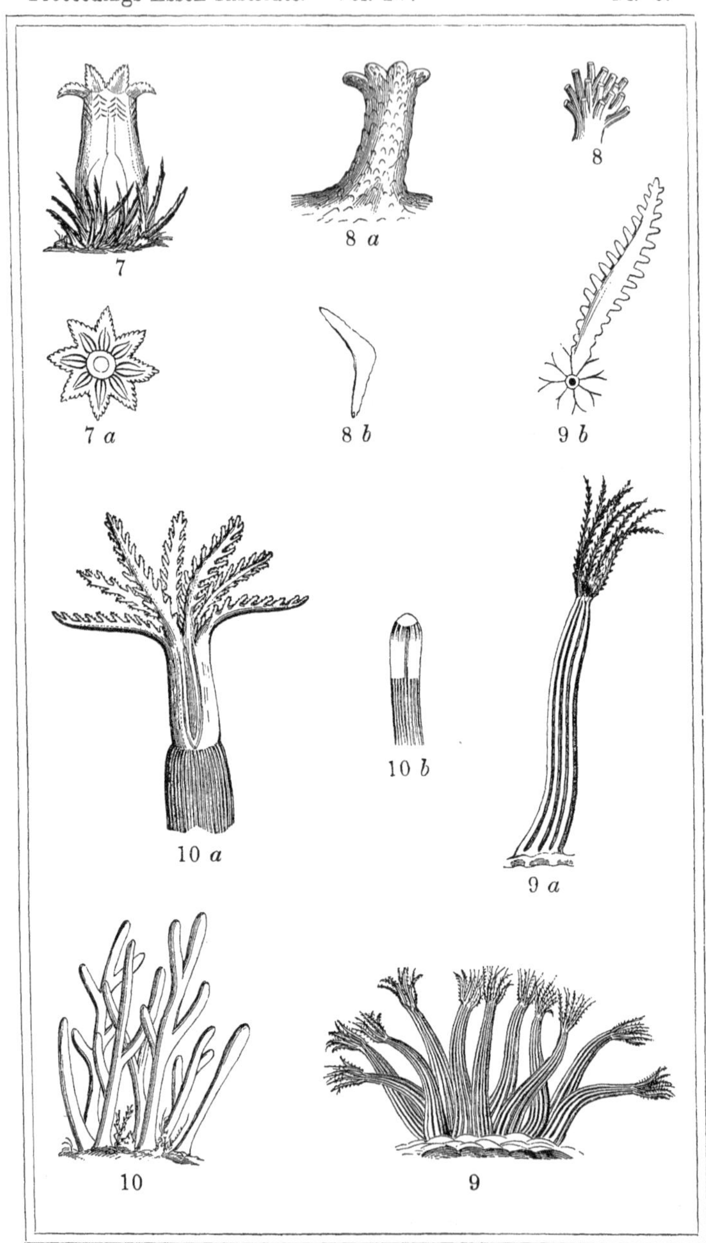

E. S. Morse, Del. A. Holland, Wood Cut Printer, Boston. L. Sanford, Eng., New Haven.

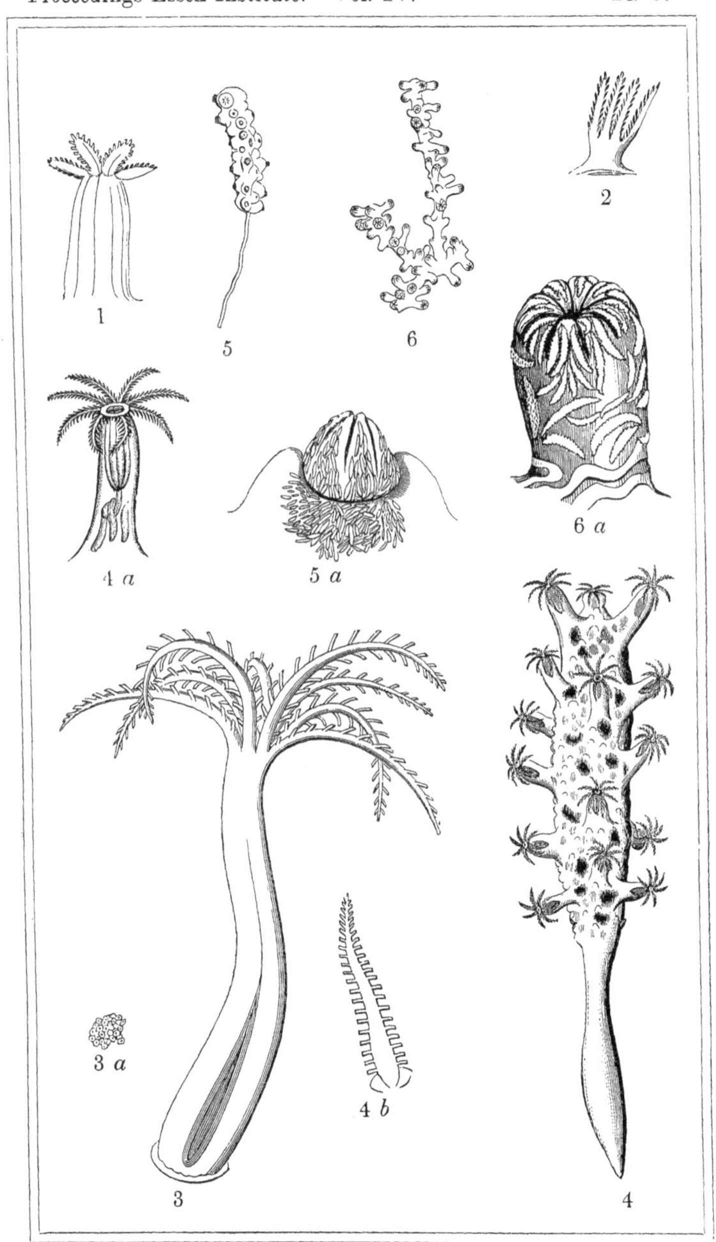

E. S. Morse, Del. A. Holland, Wood Cut Printer, Boston. L. Sanford, Eng., New Haven.

www.ingramcontent.com/pod-product-compliance
Lightning Source LLC
LaVergne TN
LVHW050524100826
845148LV00002B/431

* 9 7 8 1 4 2 5 5 2 0 0 8 3 *